# Vision Sensors - Recent Advances

*Edited by Francisco Javier Gallegos-Funes*

Published in London, United Kingdom

Vision Sensors - Recent Advances
http://dx.doi.org/10.5772/intechopen.100738
Edited by Francisco Javier Gallegos-Funes

Contributors
Libor Hargaš, Dušan Koniar, Enrique Guzmán-Ramírez, Ayax García, Esteban Guerrero-Ramírez, Antonio Orantes-Molina, Oscar Ramírez-Cárdenas, Ignacio Arroyo-Fernández, Elias Mazrooei Rad, Carola Figueroa-Flores, Ana Grasielle Dionísio Corrêa, Silvana Maria Blascovi-Assis, Natália Regina Kintschner, Thiago Leandro Liporace, Liang-Chia Chen, Sheng-Hao Huang, Bo-Han Huang, Ioannis Kypraios



First published in London, United Kingdom, 2023 by IntechOpen
IntechOpen is the global imprint of INTECHOPEN LIMITED, registered in England and Wales, registration number: 11086078, 5 Princes Gate Court, London, SW7 2QJ, United Kingdom

British Library Cataloguing-in-Publication Data
A catalogue record for this book is available from the British Library

Additional hard and PDF copies can be obtained from orders@intechopen.com

Vision Sensors - Recent Advances
Edited by Francisco Javier Gallegos-Funes
p. cm.
Print ISBN 978-1-83769-987-2
Online ISBN 978-1-83768-210-2
eBook (PDF) ISBN 978-1-83768-211-9

# Meet the editor

Francisco J. Gallegos-Funes received his Ph.D. in communications and electronics from the Instituto Politécnico Nacional de México (National Polytechnic Institute of Mexico) in 2003. He is currently an associate professor in the Escuela Superior de Ingeniería Mecánica y Eléctrica (Mechanical and Electrical Engineering Higher School) at the same institute. His areas of scientific interest are signal and image processing, filtering, steganography, segmentation, pattern recognition, biomedical signal processing, sensors, and real-time applications.

# Contents

# Preface

*Vision Sensors - Recent Advances* reflects a selection of recent developments within the field of vision sensors and their associated algorithms. Vision sensors use captured images to localize the target object's presence and orientation to the background. As the objects can appear in any area, the accuracy achieved with a single model is minimal, but overall accuracy can be improved using the combined predictions of multiple models. A bounding box is used for training these models so that the sensor can learn different object locations. Among the image processing techniques that involve the use of vision sensors are image measurement, image analysis, image transformation, image filtering, segmentation algorithms, supervised and unsupervised algorithms for classification, image recognition and parallel computing, and real-time algorithms. Recent literature reviews vision sensors with applications for localization of robot end effectors, panoramic vision sensors, wireless vision sensors, automated vision sensor inspection, multi-sensor imaging for medical data, and object recognition.

This book contains seven chapters by researchers and professionals in the field of vision sensors with applications in image processing. Chapter 1 discusses medical applications for an RGB-D multi-sensor imaging system, with particular reference to a 3D optical (RGB-D) craniofacial scanning system using multiple depth camera sensors. The system is designed to obtain an extensive dataset of scans (head and face) from various views as an alternative pre-diagnostic method for obstructive sleep apnea.

Chapter 2 considers a multi-object recognition system, using a feature descriptor and neural classifier, and based on a histogram of oriented gradients and multilayer perceptron (ORS HOG-MLP). The proposed improvement in the calculus of the HOG algorithm accurately represents different objects in multiclass applications, reducing the rate of false positives.

Chapter 3 introduces a new method for the diagnosis of Alzheimer's disease combining electroencephalogram (EEG) signals and magnetic resonance imaging (MRI). Support-vector machine and Elman neural network classifiers are used, with the optimal combined features extracted by analysis of variance to provide a good classification of the disease.

Chapter 4 proposes a saliency subitizing process (SSP) model which generates an initial saliency map using subitizing information without any seeds from unsupervised methods. This task is inspired by the ability of people to quickly and accurately identify the number of items within the subitizing range, so that the subitizing information will tell us the number of featured objects in a given image.

Chapter 5 analyzes the therapeutic effects of using a leap motion controller (LMC) sensor to assess hand fine motor dexterity, with particular reference to the box and blocks test (BBT). The LMC device and its forms of interaction in virtual reality environments (immersive and non-immersive) are examined, together with manual function assessment, traditional BBT and its virtual version (VBBT), and VBBT integrated with the LMC in physical therapy practice.

Chapter 6 introduces a method for detecting the precise 6-degrees-of-freedom (6DOF) localization of robotic arms' end effectors without referencing any additional feature target during in-line robot operation. This method facilitates precise robot positioning and monitoring.

Chapter 7 discusses a new hybrid digital-optical architecture that combines an optical correlator with a supervised and/or unsupervised classifier learning algorithm to create a fast SDF (synthetic discriminant function) k-means classifier.

I thank the researchers and professionals whose chapters have contributed to the publication of this book, and I hope that readers will find it interesting.

**Francisco Javier Gallegos-Funes**
Instituto Politécnico Nacional de México,
Mexico City, Mexico

Chapter 1

# Usage of RGB-D Multi-Sensor Imaging System for Medical Applications

*Libor Hargaš and Dušan Koniar*


## Abstract

This chapter presents an inclusion of 3D optical (RGB-D) sensors into medical clinical practice, as an alternative to the conventional imaging and diagnostic methods, which are expensive in many aspects. It focuses on obstructive sleep apnea, the respiratory syndrome that occurs in an increasing proportion of the population, including children. We introduce the novel application, a response to the request for an alternative pre-diagnostic method for obstructive sleep apnea in the region of Slovakia. The main objective of the proposed system is to obtain an extensive dataset of scans (head and face) from various views and add detailed information about patient. The application consists of the 3D craniofacial scanning system using multiple depth camera sensors. Several technologies are presented with the proposed methodology for their comprehensive comparison based on depth sensing and evaluation of their suitability for parallel multi-view scanning (mutual interference, noise parameters). The application also includes the assistance algorithm guaranteeing the patient's head positioning, graphical interface for scanning management, and standardized EU medical sleep questionnaire. Compared to polysomnography, which is the golden standard for this diagnostics, the needed data acquisition time is reduced significantly, the same with the price and accessibility.


**Keywords:** RGB-D sensors, multi-sensor system, 3D imaging, medical applications, obstructive sleep apnea, time of flight sensor, structured light sensor, stereo vision

## 1. Introduction

Obstructive sleep apnea syndrome (OSAS) is a common sleep disorder with arising prevalence. The course of the disease is followed by repeating breath interruptions during sleep. The reason is the collapse of soft upper airway tissue. This restriction of ventilation results in several breathing difficulties, such as snoring during sleep and hypoxemia. OSAS also leads to long-term changes in autonomous functions, hypertension, or reduced left ventricular function. As a result of the propagation and activation of inflammatory pathways, immune regulation is often disrupted in pediatric patients. Early diagnostics and relevant treatment are the keys for improve the health status of patients.

Polysomnography (PSG) is the golden standard for OSAS diagnostics [1]. PSG is performed for whole night in specialized sleep laboratories and the PSG device is continuously monitoring selected vital functions, such as ECG, EEG, thoraco-abdominal movements, nasal airflow, blood oxygen saturation, snoring, or body position. OSAS is confirmed if the number of apnea episodes is higher than 15 during a night and the episode takes more than 10 seconds [1, 2]. The apnea-hypopnea index (AHI) is calculated, which is used to indicate the severity of the disease.

Diagnostics and care of patients with obstructive sleep apnea vary by country and depend on the patient's symptoms. Available data suggest that most cases remain undiagnosed and untreated even in developed countries, which can increase the risk of cardiovascular, metabolic, and neural diseases and affect the quality of life. Generally, PSG is a time-consuming procedure carried out using specialized equipment, so there will be always a patient limit for undergoing the diagnostic test. As an example, there is only one specialized child and adolescent sleep laboratory in Slovakia for approximately one million children under 19 years.

Therefore, there is a reasoned demand for alternative or pre-diagnostic testing that will distinguish the patients with a high risk of OSA. Today's PSG testing can be also proved by telemedicine. Although it holds great promise to change health care delivery, it has not been proven to have the same accuracy as conventional PSG. Besides the conventional PSG, there are several supplementary testing methods. One of the best known is the sleep questionnaire [3–5] focused on the medical history of the patient and the physical examination. Another supportive diagnostic tool is the pulse oximetry or examination of a specific protein in blood serum. Mentioned questionnaires evaluate the subjective and objective symptoms, as well as the craniofacial and intraoral anatomy. These structures are considered an important indicator of the predisposition of the OSAS [6]. Nowadays there is an effort to make diagnostics more available, therefore, the emphasis is placed on the use of fast imaging techniques.

Many recent publications [7–9] focus on the face anatomy (craniofacial anthropometry, structure of the soft tissue in oral cavity, and anterior neck subcutaneous fat tissue thickness) and use advanced imaging techniques such as X-ray [10], MRI [11], and CT [12, 13]. Although, mentioned modalities do not match the criteria for cost reduction, faster procedure, and simplification of the clinical examination. Last but not least, the speed of scanning process is very important to avoid motion artifacts in resulting 3D models, especially if the system is dedicated to pediatric medicine.

As an alternative, we present an optical depth multi-sensor system that can be used excluding other emerging disadvantages with lower quality of an output model. Optical depth sensors allow capturing the nature of craniofacial anatomy needed for prediction of OSAS, such as shape and contour in a faster, cheaper, and more readily available way, compared with the other imaging techniques [14]. The geometrical precision of an output model is the key attribute for the desired application. It is also the goal of the proposed multi-camera parallel scanning system – to reconstruct a complete 3D model of the object from a collection of images taken from known camera viewpoints. Therefore, it is important to choose a suitable optical sensor with the least measurement error. For a real application, the main requirement is to obtain a complete 3D model without any noisy artifacts. In this work, we aim to evaluate each of the camera technologies: The Intel® RealSense™ Depth Camera D415 Series sensors, Stereolab's ZED Mini depth camera, Microsoft Kinect for Windows V2, and Intel® RealSense™ Camera SR300 and offer a comparison of individual operating technologies. With the selection of a suitable optical sensor, also the fact that the scanning object is a pediatric patient is taken into account.

Based on accuracy measurements, we prefer active stereo pair technology. The design, including the optimal topology of used cameras, user interface, and implementation of conventional OSAS screening questionnaire is introduced. Our effort is to predict the probability of OSAS occurrence without the need for traditional polysomnography testing. For this purpose, in the first stage of research, we use the scanning system primarily to obtain the 3D models of the patient's head, and subsequently, the database of point cloud models will be created for further research (automated extraction of key points in the face and head and automated measurement of geometric dependencies indicating the risk of OSAS). Currently, the absence of the database of 3D scans is the crucial limitation of OSAS data processing and assessment. Many studies dedicated to automated diagnostics of OSAS suffer from the datasets with small numbers of images and models. For this reason, building a huge datasets of 3D scans taken from various points of view with additive information about the patient is one of the main objectives of our research.

Additive information is a de facto electronic version of an internationally standardized sleep questionnaire. This dataset will be used for further automated diagnostics and research in this field. The result of our work is the system that consists of a fixing stand (that allows changing the camera layout) and software web-based application (includes the data annotation and the assistance support system) that helps the operator to set the patient's head into the normalized position. Using the advantage of machine learning it seems to be possible to evaluate the presence of OSAS from the point cloud representation of the patient's head and neck [15]. In the future, we assume the use of obtained dataset (composed of different views and facial expressions) with additive information in OSAS automated diagnostics. The experimental system is located in Martin University Hospital in Slovakia, Clinic of Children and Adolescents.

## 2. Related research in the field

Nowadays, the research of the new predictive diagnostic method of OSAS based on the 2D or 3D craniofacial image of the patient uses the advantage of machine learning, artificial intelligence, or statistical analysis. This research is based on the fact that OSAS occurrence is correlated with many diseases and syndromes (obesity, Down syndrome, adenotonsillar hypertrophy ... ) manifesting on the head and face. Many modern diagnostic approaches for OSAS use automated detection of selected points on the head and face (e.g. eye corners, lips, earlobes, chin ... ) and measure distance between selected points (**Figure 1**). Description of head and face with given distances serves as basic for classification process to compare with normal (physiological) model. Craniofacial points and their distances correspond with metrics obtained from paper sleep questionnaires dedicated to computing the score of OSAS risk.

Frontal and profile 2D facial photographic images of the control and experimental group are used in Ref. [16], and the features and landmarks were identified. The features were processed by support vector machine (SVM) classifier and get the resulting accuracy of 80% correct OSAS detections. In Ref [17], the authors use the training set of 3D images of 400 patients. All of them were identified with PSG, AHI index and were divided into 4 groups. The landmarks were identified manually and were expressed as the Euclidian and geodetic distance between them. The distances were considered as the features for further OSAS classification. In Ref [18], for OSAS distinction geometrical morphometry is used, also via 3D photography; in Ref [14]

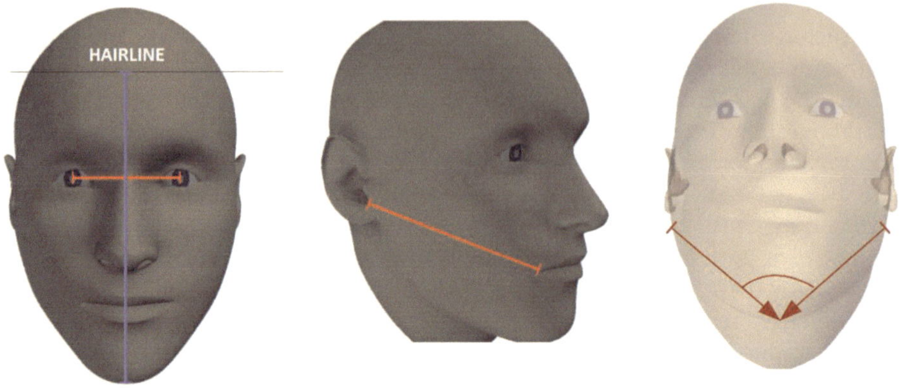


**Figure 1.**
*Selected examples of craniofacial measurement.*

convolutional neural network is a tool for OSAS prediction. Although the network was not trained on the depth data, the pretrained network achieved the 67% accuracy. Taking into account the information from previous research in this area, we can say that 3D model of the face and neck of the patient contains sufficient shape and structural features to determine the OSAS prediction. In most cases, the studies work with limited datasets, small groups of patients, or use only the frontal 3D scan of patients. As an alternative, we offer a standardized and well-described 3D model acquisition scanning system, applied in the clinical environment.

## 3. RGB-D sensors

Solving image processing tasks in various fields of research [19–21] is often helpful to obtain depth information for a better description of the scene in addition to color information. The goal is to capture the geometrical nature of the real-world object and convert it to the digital format with the highest possible accuracy. For obtaining mentioned information the depth sensors (RGB-D sensors) are widely used. They can convert the scene to the 2D plane called depth map. The depth map can be converted back to 3D space using reversed reconstruction. The depth map is usually represented by a monochromatic image, where the intensity value of the pixel represents the distance of the corresponding point from the imaging sensor. Using a combination of depth maps and color RGB images we can create a textured 3D model of scene. One of the novel application areas is the reconstruction of the 3D surface in medical research (scanning of the human heads, faces, or other body parts), taking the advantage of the noninvasive nature of digital imaging. A geometrically accurate model of head is applicable in medicine for predicting various diseases, e.g., respiratory syndromes, where the 3D representation of the patient's head and neck offers detailed visualization of craniofacial parameters with a given accuracy. In the field of 3D imaging, we know many principles and methods, e.g., photogrammetry or laser scanning devices. These methods provide high-quality 3D information; on the other side, their application is limited by the size of scanned object, size of scanner, or both object and scanner, these devices are often expensive and scanning time is too high. Also in most cases, laser scanners are not eye-safe. In the next sections, the basic principles of RGB-D sensors will be described.

### 3.1 Time of flight RGB-D sensor

The time of flight (ToF) RGB-D sensors are optical sensors that measure the depth of scene using an active light source. This light source emits an amplitude modulated signal. The emitted signal can be continuous or impulse. Most ToF cameras generate amplitude-modulated continuous waves (AMCW) with a frequency near IR for illuminating the scene [22]. The depth of scene is based on measurement of the amplitude of phase shift between received and transmitted modulated signal. The depth information for each pixel can be calculated by the synchronous demodulation of the received modulated light in the detector. The demodulation can be performed by interleaving with the original modulated signal.

### 3.2 Stereo RGB-D sensor

#### *3.2.1 Passive stereo sensor*

Passive stereo vision RGB-D sensors reproduce the depth of the scene the same way as the binocular vision of a human. The scene must be captured from different points of view. This is done by using two RGB sensors (corresponding to human eyes) horizontally separated by known distance. This distance is called baseline. For example, the depth sensor ZED MINI used in our experiments has two RGB sensors separated by a 12 cm baseline. The depth of scene is then computed based on disparity of corresponding points in single views. Solving the correspondence problem means giving a point in the image and finding the same point in another image [23]. Stereo sensors use computationally intensive algorithms to search for point matches and for computation of depth. These sensors are suitable for environments with good lighting conditions including outdoors.

#### *3.2.2 Active stereo sensor*

If the scene contains fewer color and intensity variations, or lighting conditions are not good, the passive stereo vision system can be less effective and accurate. The typical example of such environment is the texture-less surface like indoor dimly lit white walls. Active stereo vision relies on the addition of an optical projector that overlays the observed scene with a semi-random texture that facilitates finding correspondences. The current generation of RealSense D4xx sensors working in bright environments captures the texture of objects in really slight details and they are applicable also outdoors. In the case of scanning dynamic objects using the multi-sensor system, there is no limitation on how many sensors are used in a given physical layout. It does not decrease the quality of scanning process if several sensors project their light patterns to the same part of scene. All additional projectors actually improve the overall performance by adding more light and more texture [24].

### 3.3 Structured light RGB-D sensor

Depth sensors based on structured light (SLS) need additive light source. This source projects the regular patterns to the scene. The surface of the object distorts this regular pattern. If the structure of light pattern is known, the depth image of the scene can be easily computed based on its distortion [23]. Light patterns are emitted in infrared band, so the entire process is invisible to the user. If stripes are used as regular patterns, the optical resolution of the depth map can be increased by the reduction of strips width.

## 4. Measuring of accuracy of RGB-D sensors

In this section, the methods and basic measurements are described leading to proper RGB-D sensor selection, that allows capturing objects in the parallel multi-view system. Parallel multi-sensor (multi-view) system is required to reduce the scanning time (because object of interest is a pediatric patient and its potential motion can cause artifacts in resulting 3D model). Parallel means that all sensors in a given topology are capturing the object at the same time. The partial views (depth maps or point clouds produced by single sensors) must be then registered (aligned) and joined into final 3D model.

In our previous study [25], we described interference artifacts, which occur while scanning using ToF sensors. The interference is present also using several SLS sensors: projected patterns are overlaid on the surface of objects. A passive or active stereo camera pair is the technology that, in principle, does not suffer from interference in parallel multi-view systems. The depth scanning precision of sensors is compared in several recent works [26]. Known methodologies for error estimation of sensors often use a precise object and its digital model as ground truth, which is difficult to obtain. The main benefit of versatile methods described in this section is a comprehensive comparison of all sensor technologies. The measurement is based on capturing testing patterns on surfaces at small distances. The ToF sensors seem to be more accurate against the passive or active stereo pairs, according to the recent works [27, 28]. The next contribution of this research to the practice is the evaluation of the differences between these technologies in terms of accuracy. The results should show if the stereo pairs can achieve similar depth-sensing accuracy as ToF or structured light sensors at small distances.

### 4.1 The noise measurement

The noise measurement is a simple method based on evaluation of time variability of single points in the depth map. The depth is measured against a flat surface at several given distances. Depth variability (or standard deviation of the depth) in given points is represented as the noise of a depth sensor. Obviously, the noise increases with the sensor-to-object distance. All sensors were placed at distances of 0.5 m, 0.7 m, and 1 m from the flat surface. The scene was captured in 10 seconds.

### 4.2 Ideal cloud fitting

Another metric for depth error estimation is a simplified technique based on the methodology of the study [29]. Study [30] brings another technique that can be used as a generalized method for depth error estimation for any device.

The depth error estimation is based on comparing two point clouds. First point cloud is created from captured depth map and the second is the ideal software model, as shown in **Figure 2**. The results of the following measurements are extracted from our study [31]. The testing pattern is the chessboard 9 × 7 squares (square side length is 36 mm). The ideal reference point cloud was generated with the same dimensions.

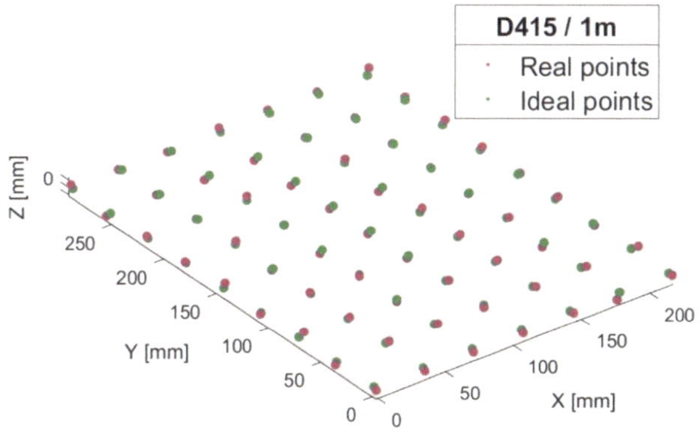


**Figure 2.**
*Real point cloud captured by RealSense D415 sensor compared with ideal one.*

The corners of a chessboard in RGB image were detected using the OpenCV algorithm. Based on equations of the pinhole camera model for projection from image coordinate system to world coordinate system $(X, Y, Z)$ we obtain the real point cloud. The $Z$-coordinate is obtained from the depth map at pixel position $(u, v)$. Following the Eq. (1) and (2), it is needed to know the intrinsic parameters $Cx$, $Cy$, $fx$, and $fy$ of cameras for getting world coordinates $X$ and $Y$:

$$X = \frac{u - C_x}{f_x} Z, \tag{1}$$

$$Y = \frac{v - C_y}{f_y} Z, \tag{2}$$

Captured (real) point cloud is fitted to the ideal one using translation and rotation estimation. As the global registration technique, Coherent Point Drift was used and final precise registration was Iterative Closest Point (ICP). The Root Mean Square Error (RMSE) of Euclidean distance was used as a relevant metric for sensor accuracy assessment:

$$RMSE = \sqrt{\frac{1}{N} \sum\nolimits_{n=1}^{N} (p_i - p_i')^2}, \tag{3}$$

where the $p_i$ and $p_i$', are the sets of coordinates of the real and ideal points, respectively.

Since we are not able to construct the precise 3D object with chessboard patterns and its ideal software model, we decide to use a flat surface. To simulate 3D scene, we captured the flat chessboard from 3 different views (**Figure 3**). The resulting error is represented as the mean value of errors obtained from views A, B, and C.

### 4.3 Ideal plane fitting

As described in the study [23], another way of depth error estimation is fitting the captured real point cloud to an ideal surface. We used the plane without chessboard captured from different positions similarly in previous method. The mean Euclidean

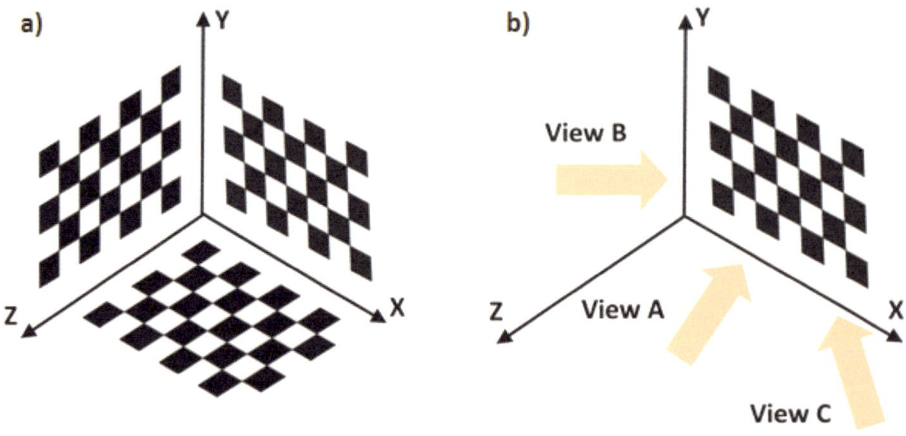


**Figure 3.**
*The possible approach on how to capture the test chessboard pattern in 3D: (a) precise 3D construction used in studies [29, 30] (b) our approach: Capturing test chessboard from several views.*

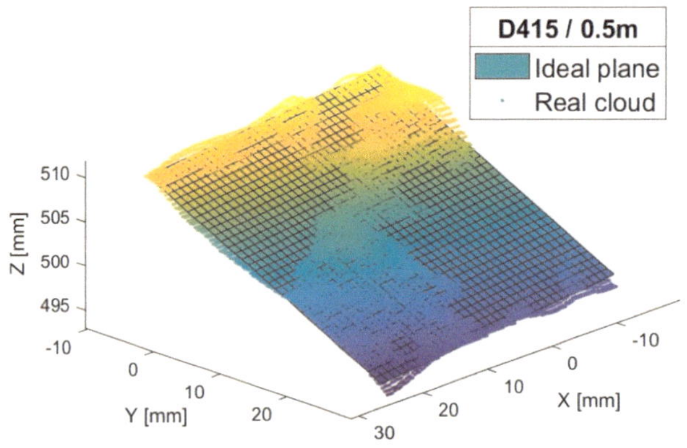


**Figure 4.**
*Real point cloud captured by RealSense D415 and fitted to ideal plane.*

distance between the ideal plane and real point cloud represent the estimated error. The fitting of real and ideal point clouds is shown in **Figure 4** [31].

## 4.4 Comparison of selected RGB-D sensors

In accuracy measurement of sensors described above, we used 4 sensors of different principles. We also assessed the suitability of these sensors in multi-sensor parallel configurations. The ToF sensor in KinectV2 and the structured light sensor in RealSense SR300 use infrared light, so their usage in parallel multi-view system is complicated due to mutual interference. The good candidates for desired imaging system are ZED MINI and RealSense D415. These sensors represent passive and active stereo pairs technology, respectively. The main depth sensor parameters of each camera are summarized in **Table 1** and available on product websites [32, 33] and comparison table in [23].

| Sensor | Technology | DFOV* | Max. resolution | FR* [fps] | Range [m] |
|---|---|---|---|---|---|
| RealSense D415 | Active stereo | 72 | 1280 × 720 | 90 | 0.3–10 |
| Kinect V2 | ToF | 70 × 60 | 512 × 424 | 30 | 0.5–4.5 |
| ZED MINI | Stereo | 110 | 4416 × 1242 | 100 | 0.15–12 |
| RealSense SR300 | Structured light | 90 | 640 × 480 | 60 | 0.2–1.5 |

**Maximal FR value might depend on resolution used/ *DFOV – Diagonal Field of View.*

**Table 1.**
*Depth sensors parameters comparison.*

| Sensor | σ for distance [mm] | | |
|---|---|---|---|
| | 500 | 700 | 1000 |
| RealSense D415 | 0.307 | 0.639 | 1.303 |
| ZED MINI | 1.499 | 1.343 | 2.180 |
| Kinect V2 | 1.151 | 1.267 | 1.375 |
| RealSense SR300 | 0.124 | 0.253 | 0.716 |

**Table 2.**
*Standard deviation of depth error for different distances.*

**Table 1** brings the comparison of key parameters of each sensor, such as resolution, diagonal field of view (DFOV), frame rate, and range for an optimal distance of sensor from object.

### *4.4.1 Comparison based on noise measurement*

The noise measurement methodology is described above and results are taken from our study [31]. The standard deviation of depth error for different distances represents amount of sensor noise. The comparison of sensors is in **Table 2**.

**Figure 5** shows the statistical comparison for a distance of 0.5 m. The amount of noise increases with the distance between the sensor and the surface. In this comparison, the offset of sensor is ignored, and also variable part of signal is taken into the account.

As seen in **Table 2**, SLS technology (RealSense RS 300) achieved the best results. As expected, there is an evident difference between ZED MINI and RealSense D415. As expected the accuracy of active stereo sensor is more accurate than passive sensor.

### *4.4.2 Comparison based on ideal cloud fitting*

In this experiment, the same sensor parameters were set, as in the previous measurement. In the case of ZED MINI and RealSense D415, the chessboard pattern was captured from distances of 0.5 m and 1 m. The results including **Table 3** are obtained from our study [31].

In this experiment, the results only for two sensors were shown, because of several negative facts. Because the RGB resolutions of ZED MINI and RealSense D415 sensors

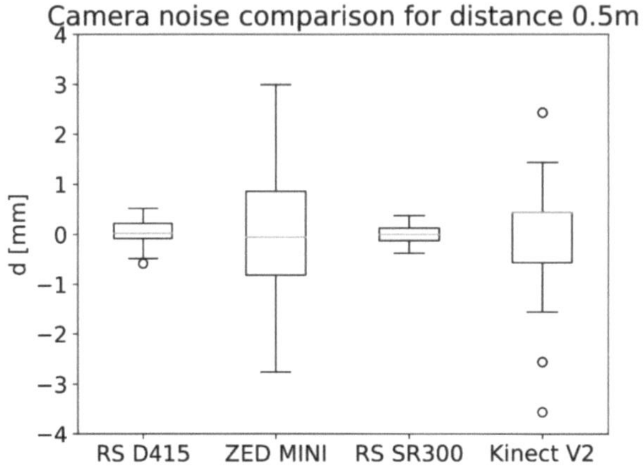


**Figure 5.**
*Noise of depth sensors comparison for distance 0.5 m.*

| Sensor | RMS error for distance [mm] | |
|---|---|---|
| | 500 | 1000 |
| RealSense D415 | 1.382 | 3.172 |
| ZED MINI | 1.803 | 4.582 |

**Table 3.**
*RMS error for multiple distances and positions using ideal cloud fitting.*

are the same, we expect the same corner detection error. For this reason, the comparison of passive and active stereo sensors in this way we consider as precise. Some problems occurred when capturing the chessboard pattern with SR300 and KinectV2 sensors. The sensor SR300 produces the depth map that contains “empty areas” of unknown depth demonstrated as black holes in depth images. Due to this fact it is not able to compute the depth of the chessboard corner point. We can say that the real point of cloud construction is impossible. Also, while using KinectV2 in a distance of more than 0.7 m, the depth map contains the black regions of unknown depth. The depth deviation is caused by different object surface reflections. Such a phenomenon, associated with ToF camera calibration is described in the study [34]. To avoid this, the color version of chessboard instead of black and white can be used. Also, the precisely constructed cube covered by the chessboard pattern is a potential way, as described in [30]. The small resolution of Kinect V2 RGB image does not allow to detect chessboard corners correctly. Due to this fact, the comparison of all 4 technologies in this way we consider inadequate.

### *4.4.3 Comparison based on ideal plane fitting*

**Table 4** taken from Ref [31] shows the comparison of all tested sensor technologies.

The estimated depth error is independent of the corners detection error, so the corresponding results in **Tables 3** and **4** are different. The difference in

| Sensor | RMS error for distance [mm] | |
|---|---|---|
| | 500 | 700 |
| RealSense D415 | 0.646 | 1.026 |
| ZED MINI | 0.782 | 1.052 |
| Kinect V2 | 1.515 | 1.588 |
| RealSense SR300 | 0.321 | 0.918 |

**Table 4.**
*RMS error for different distances and positions using ideal plane fitting.*

RMSE between structured light and active stereo pair for a distance of 0.5 m is only 0.3 mm.

## 5. Development of multi-sensor system

Based on previous measurements we decided to use active stereo sensor RealSense D415 as a key element of our imaging system. The scanning accuracy is comparable to other sensor technologies and it is absolutely sufficient for given medical use. On the other hand, active stereo pair does not suffer mutual interference in parallel mode of scanning. From the previous results, we can determine the optimal distance of head from the sensors and this distance is approximately 0.5 m. This distance is respected in a physical model of the sensor stand. For mounting 3 sensors in the fixed position, we use the constructed stand, shown in **Figure 6a**. The spatial configuration schema is shown in **Figure 6b**.

The distance $d$ is set approximately to 0.5 m. In our application, the frontal side of the object (the face) is the most important for scanning, so the layout needs to be set to obtain a high fill rate in this area. For future automated processing of captured 3D models and their normalization, the patient must sit in a normalized position. To avoid covering the important parts of the head we had to replace physical head fixation by software assistance tool mentioned later. This assistance algorithm places the head to the center of frontal sensor because the features obtained from this view (facial view) are the most important.

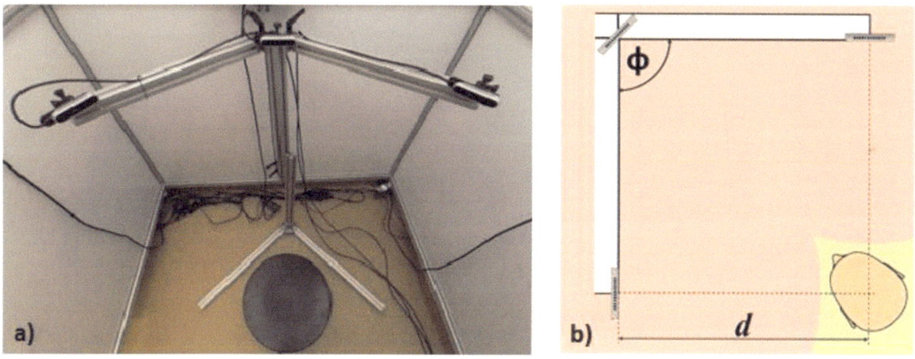


**Figure 6.**
*(a) Adjustable sensor stand. (b) Layout of sensor positions (top view).*

### 5.1 Data capturing

All the sensors are connected via the USB 3 interface to the acquisition computer, which allows capturing the color and depth frames with a 30 fps frame rate at Full HD spatial resolution. The system for the capturing and data processing is designed as GUI running as a web application. The server application is created using Python. This application acquires the depth and RGB color image frames from all sensors and streams the image data to user interface. The RealSense SDK tool is used for controlling the sensors, flask for server operation, and OpenCV framework for image processing. Image acquisition and streaming of the image data run in separated threads. The data flow is reduced due to JPEG encoding of images (depth and also RGB). When the server application receives a request for saving from the user interface, the saving procedure is triggered. Acquired images are stored locally in a temporary folder. When image acquisition process is finalized, the content of the temporary folder is zipped, encrypted, and named by patient identifier and actual time. The script then copies the ZIP to external server storage to collect and backup the data. When the external server is not accessible, the file is queued and sent in the next time [35].

### 5.2 Graphical user interface of system

The graphical user interface enables to control the scanning of the patient's head and neck and also to annotate the captured data. For annotating, the digital version of standardized sleep questionnaire is part of application and it is described later. Web-based design of this interface enables to use of any portable device in the network to provide scanning and annotating the data. Also, the interface can be accessed locally on the PC where the server runs. The main window of the interface is shown in **Figure** 7. Menu on the left side contains several settings:

- # of frames – the number of color and depth frames saved by one shot,
- ID – personal identifier of patient,
- Expression – the facial expression of the patient.

Number of frames taken by "one shot" helps to provide temporal filtering of data. Considering our study [25], resulting depth image from given view is a product of averaging the stack of images in buffer. This averaging helps to eliminate noise artifacts in 3D reconstruction.

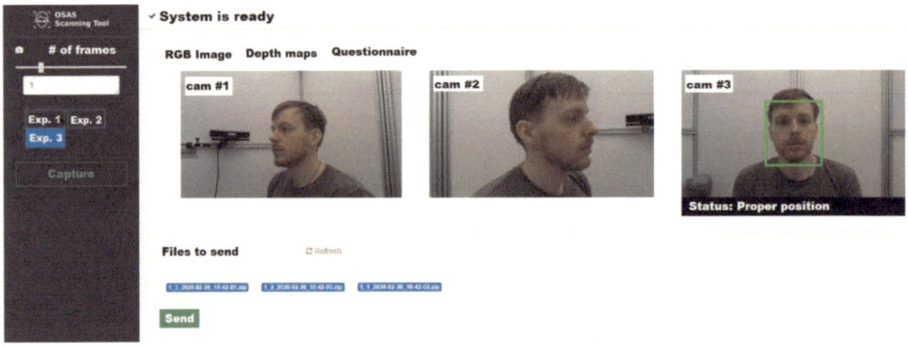


**Figure 7.**
*Graphical user interface – Main window.*

Facial expression is functionality prepared for further research. It could be interesting to correlate the OSAS detection based on normalized face expression (e.g., smile, neutral expression).

After capturing request, the data are zipped and sent to the acquisition server. If any error, the data are saved locally and resent to the acquisition server in next capturing request. If the webserver is unavailable, the warning message is displayed. The data saved on server can be identified by personal ID of patient. The user can switch between color and depth view.

### 5.3 Normalized head position assistance algorithm

For obtaining the most precise, accurate, and normalized 3D scans, we need to set the patient's head to a defined position (as equal as possible in all patients). This task may be very difficult and stressful for young patients. Based on this fact, we created the head position assistance tool based on the algorithm of eyes and head detection. Detected eye position is used to compute the difference from ideal eye position. The depth map is also available so, the algorithm can get the difference in eye positions in Z-axis. This information is obtained only from the central sensor images. In **Figure 8**, we can see the angle offsets of detected eyes from ideal position. The angles $\alpha$ and $\beta$ are used for determining how much is the head rotated or tilted. The limits were chosen empirically and they can be improved during further research. Based on the depth map, we are also able to compute the distance of the head from sensors (if it is too close or too far from sensor). In other words, head positioning assistance tool helps to keep the head in red highlighted area according to **Figure 6b** [35].

When the head position is inside the optimal range, imaging can be provided. If the position of head is outside the tolerance, the moving, rotation, and tilting commands are shown for the user on the screen to adjust the head position. For detection of face in actual sensor view, the Viola-Jones algorithm is used. This algorithm is a frequently used tool for given object detection. The original algorithm is used to detect and classify the objects into several classes. In our case, it was trained for human faces. In comparison with other algorithms, the training time is relatively high, but detection is very fast. The algorithm uses Haar basis feature filters and it does not use multiplications [36]. The computation time is minimized by placing the classifiers with fewest

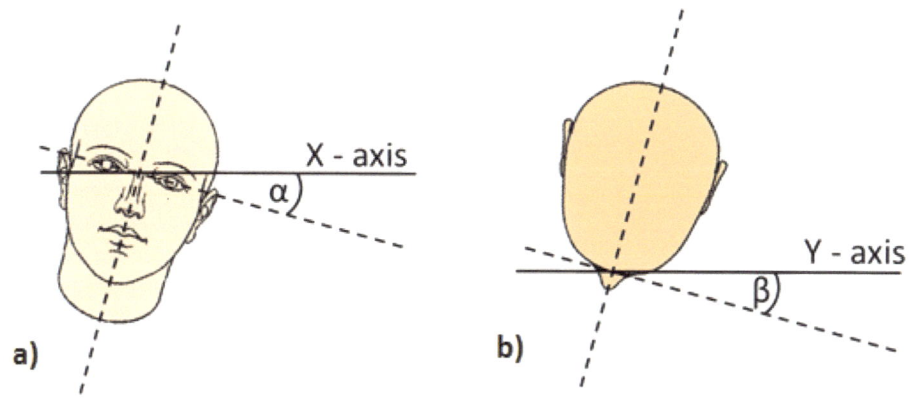


**Figure 8.**
*Difference of eye positions: (a) X-axis offset. (b) Y-axis offset.*

features at the beginning of cascade. The features are most commonly trained using Ada-Boost algorithm. This method selects only those features that improve the detection accuracy and potentially decrease the execution time.

### 5.4 The annotation questionnaire

In addition to mentioned functions, the application includes an online questionnaire, which is the digital format of the EU questionnaire [5]. Besides the 3D imaging of patient heads, the specialist (user) is able to insert additional information about the patient (age, weight, subjective rating of intraoral anatomy ... ) [35]. This additive information can be used to extend features for machine learning methods and automated diagnostics based on artificial intelligence. Implemented electronic questionnaire is shown in **Figure 9**.

### 5.5 Experimental results

After pilot testing, the system was placed in an experimental workplace inside the sleep laboratory of Clinic of Children and Adolescents in University Hospital Martin, Slovakia.

Application is designed in a simple layout and also assists medical staff (as we can see as an example in **Figure 10**) to obtain the best results without any sophisticated manipulation.

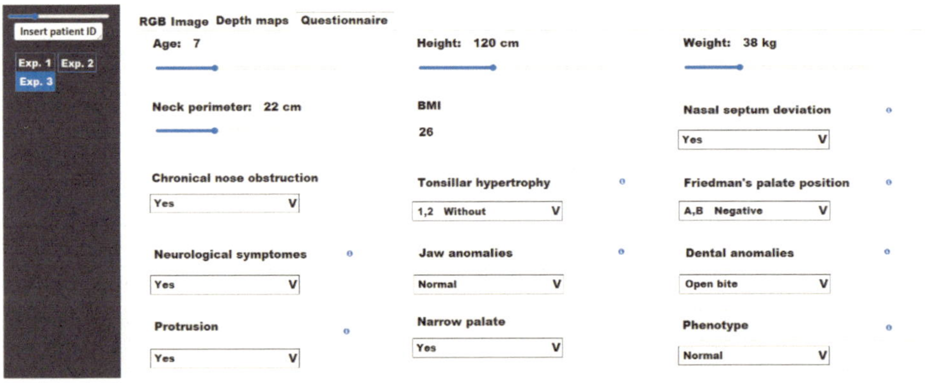


**Figure 9.**
*Graphical user interface – online EU questionnaire (excerpt).*

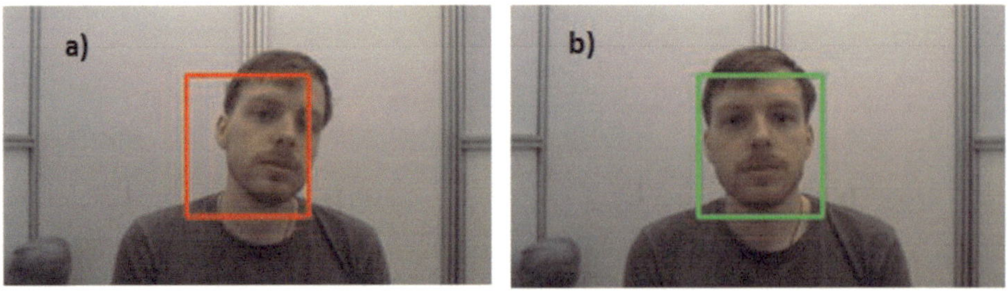


**Figure 10.**
*Head positioning assistance: (a) incorrect head position; the command is tilt head right. (b) Correct position of head.*

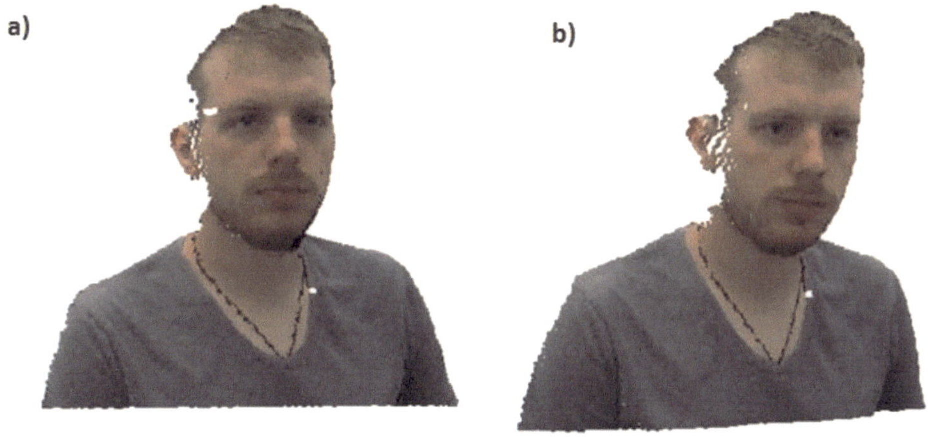


**Figure 11.**
*The resulting 3D model of patient: (a) frontal view. (b) Rotated view.*

After capturing the stack of images from single sensors, the 3D reconstruction of data can be provided. Using intrinsic camera parameters, we can compute the colored point cloud from depth and color frame. The point clouds of individual views must be also denoised and transformed into a common coordinate system. As a denoising method, the statistical outlier filter was used. The resulting 3D model with the possibility of rotation is shown in **Figure 11**.

The partial point clouds from single sensors are registered by the global registration RANSAC method and local refinement is done by ICP (Iterative Closest Point) algorithm. For further scientific research, it is very interesting to implement and compare different filtration algorithms for depth maps or point clouds, registration methods, and calibration algorithms that can improve the accuracy of models. Nowadays, it exists a lot of machine learning methods that can obtain the relevant features from the head (from depth maps, RGB images, or directly from 3D models) and will finalize the feature vector for automated diagnostics.

## 6. Conclusion

Our study focuses on the development of the multi-sensor scanning system, that aims to be the future pre-diagnostic tool for obstructive sleep apnea diagnostics. Because the obstructive sleep apnea syndrome can correspond with some abnormalities in cranio-facial parameters on the head, 3D scanning of head can be a promising procedure to obtain an automated method for OSAS screening. A system for early screening can easily prioritize the patients for complex diagnostics and then for early therapy. Especially in Slovakia, the waiting periods for conventional OSAS diagnostic can be several months.

RGB-D sensors are relatively non-expensive sensors with increasing popularity used in many fields: from entertainment to mechanical engineering or medical applications. To complete the 3D scanning system for biomedical use, the main research was focused on the selection of suitable RGB-D sensor for obtaining the accurate model of the head and neck. This model can be used for noninvasive automated procedures. After selecting the representative for all technologies of RGB-D sensors we used some metrics, which can compare their accuracy. The noise measurement, the ideal point cloud fitting, and ideal plane fitting were selected for this assessment.

After the series of experiments, we can say that the difference in accuracy between the all sensors is not so significant and all of them could be used for our implementation. On the other hand, considering the second condition – multi-sensor parallel system – the mutual interference of sensors must be taken into the account. Because ToF sensors and also SLS sensors interfere and can generate interference artifacts, we focused on stereo pair technology of RGB-D sensors. Finally, we selected the active stereo pair Intel RealSense D415. Based on depth error, the optimal distance of the sensor from object is set to 0.5 m. The system with 3 sensors respects this distance.

The scanning system is driven by a web-based application with simple graphical user interface. 3D scans can be extended by information from a digitized EU sleep questionnaire. The database of 3D models with information form questionnaire is strictly needed to build an automated diagnostic system based on machine learning or artificial intelligence. These methods are now state of the art in many imaging and signal processing tasks. System is now implemented in clinical environment to obtain first elements of the dataset.

Further research can be oriented for selecting and implementing the filtration methods for obtained data, registration methods for partial models from single sensors, and calibration algorithms for the case of changes in sensor layout.

## Acknowledgements

Results of this research are supported by grant no. APVV-15-0462: Research on sophisticated methods for analyzing the dynamic properties of respiratory epithelium's microscopic elements and grant no. APVV-17-0218: Investigation of biological tissues with electromagnetic field interaction and its application in the development of new procedures in the design of electrosurgical instruments.

Special thanks go to the medical experts and employees from the Clinic of Children and Adolescents (Jessenius Faculty of Medicine in Martin and Martin University Hospital).

## Conflict of interest

The authors declare no conflict of interest.

## Author details

Libor Hargaš* and Dušan Koniar
Faculty of Electrical Engineering and Information Technology, Department of Mechatronics and Electronics, University of Zilina, Zilina, Slovakia

*Address all correspondence to: libor.hargas@feit.uniza.sk

## References


[1] Ceska R. Interna. Triton; 2015. ISBN: 978-80-7387-885-6

[2] Villa MP, Pietropaoli N, Supino MC, Vitelli O, Rabasco J, Evangelisti M, et al. Diagnosis of pediatric obstructive sleep apnea syndrome in settings with limited resources. JAMA Otolaryngology–Head & Neck. 2015;**141**:990. DOI: 10.1001/jamaoto.2015.2354

[3] Netzer NC, Stoohs RA, Netzer CM, Clark K, Strohl KP. Using the Berlin questionnaire to identify patients at risk for the sleep apnea syndrome. Annals of Internal Medicine. 1999;**131**:485. DOI: 10.7326/0003-4819-131-7-199910050-00002

[4] Johns MW. A new method for measuring daytime sleepiness: The Epworth sleepiness scale. Sleep. 1991;**14**: 540-545. DOI: 10.1093/sleep/14.6.540

[5] Feketeová E, Mucska I, Klobučníková K, Grešová S, Stimmelová J, Paraničová I, et al. EU questionnaire to screen for obstructive sleep Apnoea validated in Slovakia. Central European Journal of Public Health. 2018;**26**:S32-S36. DOI: 10.21101/cejph.a5278

[6] Myers KA, Mrkobrada M, Simel DL. Does this patient have obstructive sleep apnea?: The rational clinical Examina-tion systematic review. JAMA. 2013;**310**:731. DOI: 10.1001/jama.2013.276185

[7] Miles PG, Vig PS, Weyant RJ, Forrest TD, Rockette HE. Craniofacial structure and obstructive sleep apnea syndrome — A qualitative analysis and meta-analysis of the literature. American Journal of Orthodontics and Dentofacial Orthopedics. 1996;**109**:163-172. DOI: 10.1016/S0889-5406(96)70177-4

[8] Capistrano A, Cordeiro A, Capelozza Filho L, Almeida VC, de Silva PIC, Martinez S, de Almeida- Pedrin RR. Facial morphology and obstructive sleep apnea. Dentofacial Press Journal of Orthodontics. 2015;20:60–67. DOI: 10.1590/2177-6709.20.6.060-067.oar

[9] Hoekema A, Hovinga B, Stegenga B, De Bont LGM. Craniofacial morphology and obstructive sleep Apnoea: A Ceph-alometric analysis. Journal of Oral Rehabilitation. 2003;**30**:690-696. DOI: 10.1046/j.1365-2842.2003.01130.x

[10] de Mello Junior CF, Guimarães Filho HA, de Gomes CAB, de Paiva CCA. Radiological findings in patients with obstructive sleep apnea. Jornal Brasileiro de Pneumologia Publicacao Officials Society Brasilian Pneumology E Tisilogia. 2013;39:98–101. DOI:10.1590/s1806-37132013000100014

[11] Butorova E, Elfimova E, Shariya M, Litvin A. MRI measurement of airway soft tissues parameters in patients with obstructive sleep Apnoe. Journal of Hypertension. 2016;**34**:e331. DOI: 10.1097/01.hjh.0000492316.06821.77

[12] Chousangsuntorn K, Bhongmakapat T, Apirakkittikul N, Sungkarat W, Supakul N, Laothamatas J. Computed to-mography characterization and comparison with polysomnography for obstructive sleep apnea evaluation. Journal of Oral and Maxillofacial Surgery. 2018;**76**:854-872. DOI: 10.1016/j.joms.2017.09.006

[13] Barkdull GC, Kohl CA, Patel M, Davidson TM. Computed tomography imaging of patients with obstructive sleep apnea. The Laryngoscope. 2008;

**118**:1486-1492. DOI: 10.1097/MLG.0b013e3181782706

[14] Islam SMS, Mahmood H, Al-Jumaily AA, Claxton S. Deep learning of facial depth maps for obstructive sleep apnea prediction. In: Proceedings of the 2018 International Conference on Machine Learning and Data Engineering, iCMLDE, 3-7 December 2018. Sydney, NSW, Australia: IEEE; 2018. DOI: 10.1109/iCMLDE.2018.00036

[15] Sutherland K, Lee RWW, Petocz P, Chan TO, Ng S, Hui DS, et al. Craniofacial phenotyping for prediction of obstructive sleep Apnoea in a Chinese population. Respirology Carlton Vic. 2016;**21**:1118-1125. DOI: 10.1111/resp.12792

[16] de Chazal P, Tabatabaei Balaei A, Nosrati H. Screening patients for risk of sleep apnea using facial photographs. In: Proceedings of the 2017 39th Annual International Conference of the IEEE Engineering in Medicine and Biology Society (EMBC). Seogwipo: IEEE; 2017. pp. 2006-2009

[17] Eastwood P, Gilani SZ, McArdle N, Hillman D, Walsh J, Maddison K, et al. Predicting sleep apnea from three-dimensional face photography. Journal of Clinical Sleep Medicine. 2020;**16**:493-502. DOI: 10.5664/jcsm.8246

[18] Ozdemir ST, Ercan I, Can FE, Ocakoglu G, Cetinoglu ED, Ursavas A. Three-dimensional analysis of craniofacial shape in obstructive sleep apnea syndrome using geometric Morphometrics. International Journal of Morphology. 2019;**37**:338-343. DOI: 10.4067/S0717-95022019000100338

[19] Lin T, Liu X. An intelligent recognition system for insulator string defects based on dimension correction and Opti-mized faster R-CNN. Electrical Engineering. 2020:1-9. DOI: 10.1007/s00202-020-01099-z

[20] Uribe FA, Flores J. Parameter estimation of arbitrary-shape electrical cables through an image processing technique. Electrical Engineering. 2018;**100**:1749-1759. DOI: 10.1007/s00202-017-0651-y

[21] Yan Z, Shi B, Sun L, Xiao J. Surface defect detection of aluminum alloy welds with 3D depth image and 2D gray image. International Journal of Advanced Manufacturing Technology. 2020;**110**:741-752. DOI: 10.1007/s00170-020-05882-x

[22] Bulczak D, Lambers M, Kolb A. Quantified, interactive simulation of AMCW ToF camera including multipath effects. Sensors. 2018;**18**:13. DOI: 10.3390/s18010013

[23] Giancola S, Valenti M, Sala R. A Survey on 3D Cameras: Metrological Comparison of Time-of-Flight, Structured-Light and Active Stereoscopy Technologies. Switzerland AG: Springer Nature; 2018. DOI: 10.1007/978-3-319-91761-0. ISBN 978-3-319-91760-3

[24] Grunnet-Jepsen A, Winer P, Takagi A, Sweetser J, Zhao K, Khuong T, et al. Using the Real Sense D4xx Depth Sensors in Multi-Camera Configurations. Available from: https://simplecore.intel.com/realsensehub/wp-content/uploads/sites/63/Multiple_Camera_WhitePaper04.pdf [Accessed: August 28, 2022]

[25] Volak J, Koniar D, Jabloncik F, Hargas L, Janisova S. Interference artifacts suppression in systems with multiple depth cameras. In: Proceedings of the 2019 42nd International Conference on Telecommunications and Signal Processing, TSP, 01-03 July 2019.

Budapest, Hungary: IEEE; 2019. DOI: 10.1109/TSP.2019.8768877

[26] Langmann B, Hartmann K, Loffeld O. Depth camera technology comparison and performance evaluation. In: Proceedings of the 1st International Conference on Pattern Recognition Applications and Methods. 2012. DOI: 10.5220/0003778304380444

[27] Vit A, Shani G. Comparing RGB-D sensors for close range outdoor agricultural phenotyping. Sensors. 2018; **18**:4413. DOI: 10.3390/s18124413

[28] Chiu C-Y, Thelwell M, Senior T, Choppin S, Hart J, Wheat J. Comparison of depth cameras for three-dimensional reconstruction in medicine. Proceedings of the Institution of Mechanical Engineers, Part H: Journal of Engineering in Medicine. 2019;**233**:938-947. DOI: 10.1177/0954411919859922

[29] Ortiz LE, Cabrera VE, Goncalves LMG. Depth data error modeling of the ZED 3D vision sensor from Stereolabs. ELCVIA: Electronic Letters on Computer Vision and Image. 2018;**17**:1-15. DOI: 10.5565/rev/elcvia.1084

[30] Fernandez L, Avila V, Goncalves L. A generic approach for error estimation of depth data from (stereo and RGB-D) 3D Sensors. Preprints. 2017. DOI: 10.20944/preprints201705.0170.v1

[31] Bajzik J, Koniar D, Hargas L, Volak J, Janisova S. Depth sensor selection for specific application. In: Proceedings of the 2020 ELEKTRO, 25-28 May 2020. Taormina, Italy: IEEE; 2020. DOI: 10.1109/ELEKTRO49696.2020.9130293

[32] Intel® RealSenseTM Computer Vision - Depth and Tracking Cameras Available online: https://www.intelrealsense.com/ [Accessed: January 19, 2021]

[33] Stereolabs - Capture the World in 3D Available online: https://www.stereolabs.com/ [Accessed: January 19, 2021]

[34] Lindner M, Kolb A. Calibration of the intensity-related distance error of the PMD TOF-camera. In: Proceedings Volume 6764, Intelligent Robots and Computer Vision XXV: Algorithms, Techniques, and Active Vision, Event: Optics East. Boston, MA, United States. 2007. DOI: 10.1117/12.752808

[35] Stefunova S, Koniar D, Hargas L, Bulava J. Multi-camera scanning system for collecting and annotating 3D models of the head and neck. In: Proceedings of the International Conference on Electrical, Computer, Communications and Mechatronics Engineering, ICECCME, 07-08 October 2021, Mauritius. 2021. DOI: 10.1109/55ICECCME52200.2021.9590878

[36] Viola P, Jones M. Rapid object detection using a boosted cascade of simple features. In: Proceedings of the 2001 IEEE Computer Society Conference on Computer Vision and Pattern Recognition, CVPR, 08-14 December 2001. Kauai, HI, USA: IEEE; 2001. DOI: 10.1109/CVPR.2001.990517

**Chapter 2**

# Multi-Object Recognition Using a Feature Descriptor and Neural Classifier

*Enrique Guzmán-Ramírez, Ayax García,*
*Esteban Guerrero-Ramírez, Antonio Orantes-Molina,*
*Oscar Ramírez-Cárdenas and Ignacio Arroyo-Fernández*

## Abstract

In the field of object recognition, feature descriptors have proven to be able to provide accurate representations of objects facilitating the recognition task. In this sense, Histograms of Oriented Gradients (HOG), a descriptor that uses this approach, together with Support Vector Machines (SVM) have proven to be successful human detection methods. In this paper, we propose a scheme consisting of improved HOG and a classifier with a neural approach to producing a robust system for object recognition. The main contributions of this work are: First, we propose an improved gradient calculation that allows for better discrimination for the classifier system, which consists of performing a threshold over both the magnitude and direction of the gradients. This improvement reduces the rate of false positives. Second, although HOG is particularly suited for human detection, we demonstrate that it can be used to represent different objects accurately, and even perform well in multi-class applications. Third, we show that a classifier that uses a neuronal approach is an excellent complement to a HOG-based feature extractor. Finally, experimental results on the well-known Caltech 101 dataset illustrate the benefits of the proposed scheme.

**Keywords:** multi-object recognition systems, object representation based on feature descriptor, histogram of oriented gradients, classifier with a neural approach

## 1. Introduction

Computer vision is a discipline through which a machine is enabled in order to recognize the world around it using visual perception, allowing it to deduce the structure and properties of a three-dimensional world from one or more two-dimensional images. In this respect, Forsyth-Ponce and Ballard-Brown argue that computer vision refers to the construction of explicit and significant descriptions of physical objects from images [1, 2]. That is, computer vision enables a machine to extract and analyze spectral, spatial, and temporal information of the different objects contained within an image. While spectral information includes frequency (color) and intensity (grayscale), spatial information refers to aspects such as shape and position (one, two, and three dimensions) and temporal information comprising stationary aspects (presence and/or absence) and time-dependent (events, movements, and processes).

Due to this ability, tasks such as failure detection [3, 4]; verification [5, 6]; identification [7, 8]; tracking analysis [9, 10]; and recognition [11, 12] can be performed by a computer vision system.

The object recognition task is of particular interest to this research as it plays a significant role in a computer vision system and is necessary even in order to complete some of the tasks listed above. It is evident that there are an increasing number of areas and/or applications requiring object recognition, e.g., fruit sorting [13], face detection [14], people detection [15], face recognition [16], object tracking [17], automatic traffic sign recognition [18], and vehicle license plate recognition [7], among many others. Gonzales and Woods define object recognition as the task of organizing input data into previously defined classes, using significant features extracted from objects that are immersed in an image containing irrelevant details [19]. Considering this definition, it is evident that both feature extraction and classification are extremely important for an object recognition task to achieve its aim. The feature extraction process applies operations to an image in order to obtain information describing the objects it contains. Moreover, this information should be able to discriminate between different object classes. The goal of this process is to improve the effectiveness and efficiency of the classification process [20]. The classification process uses the information generated by the feature extractor to perform both phases comprising it, learning, (thereby creating a bank of models), and recognition (responsible for determining which objects belonging to the bank of models is present in analyzed image) [21].

The development of the work presented in this paper is motivated by the increasing demand and necessity of techniques and algorithms that efficiently perform tasks related to the recognition of objects, as well as their implementation in real applications, both industrial and research, where a visual perception system is required. Considering the above, this work proposes the design and implementation of an object recognition system using methods based on feature descriptor and neural classifier. We will focus on a type of descriptor called the feature descriptor. In their original form, these descriptors characterize shape in 2D images as histograms of edge pixels, HOG (Histograms of Oriented Gradients) algorithm belongs to this family of descriptors [22]. Two practical objectives have been defined to meet the proposal. First, study and development of an improved HOG algorithm that can accurately represent different objects. Second, to demonstrate that the performance of a neuronal approach in the task of classification and labeling of the data generated by HOG is competitive with the techniques currently used.

The HOG method for object representation, proposed by Dalal and Triggs [23], describes the appearance of local regions (objects) within an image by means of the distribution of intensity gradients or edge directions. For this purpose, the HOG method applies a similar principle to be used by different methods, such as edge orientation histograms [24], shape contexts [25], and scale-invariant feature transform (SIFT) [26], which counts the number of occurrences of gradient orientation in specific portions of an image but differs in that it is computed on a dense grid of uniformly spaced cells and uses overlapping local contrast normalization for improved accuracy.

HOG has demonstrated that it is capable of generating representations that provide discriminative information from the objects in an image using normalized representations of objects. Since it operates on local cells, it is invariant to geometric and photometric transformations. For improved accuracy, the local histograms can be contrast-normalized by calculating a measure of the intensity across a larger region of the image, called a block, and then using this value to normalize all cells within the block. This normalization results in better invariance to changes in illumination and shadowing. Furthermore, HOG is invariant to changes in the data background as well as object position.

Although, the HOG method is particularly suited for human detection in images [11, 12, 23, 27–29], in this paper we show that it can be used to represent different objects accurately, and even perform well in multi-class applications.

On the other hand, artificial neural networks (ANN) have been successfully used in a variety of classification tasks in real industrial, scientific, and business applications [13, 30–33]. Several kinds of neural networks can be used for this purpose, but we decided to use feedforward multi-layer networks or multi-layer perceptron (MLP), which are the most widely studied and used neural network classifiers.

MLP offers features that make it a competitive alternative to various conventional classification methods [34]. First, adaptability, MLP is capable of developing its own feature representation. That is to say, MLP can organize data into the vital aspects or features that enable one pattern to be distinguished from another. Second, generalization, MLP has the ability to respond appropriately to input patterns different from those involved in the learning process. Third, MLP is a universal function approximator, thus it can approximate any function with arbitrary accuracy. Fourth, since MLP uses a nonlinear activation function, it is a nonlinear model, which makes it flexible in modeling real-world complex relationships.

### 1.1 Related work

Similar works with which our proposal has been compared, because they use descriptors with similar characteristics to ours, are described below. In [35], an object model is generative and probabilistic, so appearance, scale, shape, and occlusion are all modeled by probability density functions, which here are Gaussians. Learning is carried out using the expectation–maximization algorithm, presented in [36], which iteratively converges to learn an object category using the detecting regions and their scales, and then estimating the parameters of the above densities from these regions, such that the model gives a maximum-likelihood description of the training data. Recognition proceeds by first detecting features, and then evaluating these features through a process Bayesian, using the model parameters estimated in the learning.

Zhang et al. proposed a scheme that represents the local features of the object by means of PCA-SIFT method and the global features using shape context method [37]. Both sets of features are presented to two-layer AdaBoost training network. Boosting refers to the general method of producing a very accurate prediction rule by combining relatively inaccurate rules-of-thumb. It has been used widely in computer vision, particularly for object recognition. Layer 1 chooses as "good" features, those that have the best ability to discriminate the target object class from the nontarget object class. Then, layer 2 locates the final "good" features based on the distances between the most discriminant local features selected by layer 1. This two-layered boosting method produces two strong classifiers, which can then be used in a cascaded for recognition tasks.

On the other hand, in [38] Zhang et al. presented a scheme that represents images as distributions (signatures or histograms) of features extracted with different key point detectors and descriptors. The proposed scheme represents an object from the union of a detector with a descriptor. Two complementary local region detector types are used: The Harris-Laplace detector, which responds to corner-like regions, and the

| Method | Feature extraction | Classifier |
| --- | --- | --- |
| [35] | Detector of Kadir and Brady | Bayesian approach |
| [37] | PCA-SIFT/shape context | multi-layer Adaboost |
| [38] | SIFT/SPIN/RIFT | SVM |
| [39] | Hessian-Laplace/Harris-Laplace context | k-means/Agglomerative/RNN |
| [40] | Histogram-based descriptor | AdaBoost |
| Our proposal | HOG | MLP |

**Table 1.**
*Main features summary of the mentioned schemes and our proposal.*

Laplacian detector, which extracts blob-like regions. At the most basic level, these two detectors are invariant to scale transformations only. To achieve invariance to other transformations, such as rotation or illumination, this scheme may include the descriptors SIFT, SPIN, and RIFT. For classification process, the authors use the well-known Support Vector Machines.

In [39], Leibe et al. propose a novel method for detecting and localizing objects of a visual category in cluttered real-world scenes. For the objects representation, this proposal introduces a basic object representation model known as Implicit Shape Model (ISM). ISM consists of a class-specific alphabet of local appearances that are prototypical for the object category, and of a spatial probability distribution that specifies where each codebook entry may be found on the object. Then, the object detection is implemented as a probabilistic Hough voting procedure from which hypotheses are found by a scale-adaptive Mean-Shift search.

More recently, Leptev showed how histogram-based image descriptors can be combined with a boosting classifier to provide a robust object detector [40]. Each feature is represented by a histogram of local image measurements within a region. For this purpose, HOG features are adopted and considered histograms of alternative image measurements such as color and second-order image derivatives. The HOG features are formed from orientation of local image gradient at each point using Gaussian derivatives of image computed for scale parameter defined. The histograms are normalized to the l1 unit norm. At the training, the features for normalized training images are computed, and apply AdaBoost to select a set of features and the corresponding weak classifiers optimizing classification performance.

The overall organization of the paper is as follows, after the introduction, we examine theoretical issues of HOG and MLP in Section 2. Section 3 presents the object recognition system based on HOG and MLP proposed in this paper. Experimental results are discussed in Section 4. Finally, Section 5 concludes the paper.

**Table 1** shows a summary of the main features that the mentioned schemes and the one proposed in this paper possess.

## 2. Theoretical background of HOG and MLP models

As mentioned earlier, we will focus on a type of descriptor called the feature descriptor. In general terms, this approach uses 4 steps to calculate a descriptor from an image, 1. An edge detector is applied to the image. 2. a basis point is chosen, which is a coordinate in the edge map, then a template is defined, mainly circular, centered at that point and it is divided into sections of the same size. These sections divide the image into regions, each of which corresponds to one dimension of the feature vector. 3. The value of a dimension is calculated as the number of edge pixels that fall into the region (a histogram that summarizes the spatial distribution of edges in the image relative to the chosen basis point). It is common to use the term "bin" to refer to the region in the image as well as the dimension in the feature vector. 4. Feature vector normalization.

With this approach, if the bins were small enough to each contain one pixel, then the histogram would be an exact description of the shape in the support region.

The HOG descriptor is a member of this family, it was proposed by Dalal and Triggs in [23] and extensively documented by Dalal in his PhD thesis, which was supervised by Triggs [41]. HOG offers a successful and popular object representation, particularly with human representation [22, 42]. HOG is inspired by SIFT, thus it can be regarded as a dense version of SIFT. Both algorithms are based on histograms of gradient orientations weighted by gradient magnitudes. However, there are important differences between these algorithms. First, these two algorithms differ slightly with regard to the type of spatial bins that they use, as HOG has a more sophisticated way of binning. Second, HOG computes the descriptors by means of blocks in dense grids at some single scale without orientation alignment, while in SIFT, descriptors are computed at sparse, scale-invariant key image points, and rotated to align orientation. Third, SIFT only computes the gradient histogram for patches around specific interest points obtained by taking the difference of Gaussians in the scale space (this is a local descriptor and SIFT features are usually compared by computing the Euclidean distance between them). HOG, on the other hand is computed for an entire image by dividing it into smaller cells and summing up the gradients over every pixel within each cell in an image (HOG is used to classify patches using classifiers such as SVM). Finally, SIFT is used for the identification of specific objects since the Gaussian weighting involved enables it to describe the importance of a particular point, while HOG does not have such a bias. HOG, therefore, is better suited to the classification task than SIFT.

The main idea behind the HOG descriptors is that the appearance and shape of a local object within an image can be described by the distribution of intensity gradients or edge directions. That is to say, the HOG features concentrate on the contrast of silhouette contours against the background. HOG is a window-based descriptor, whereby the window is typically computed by dense sampling over all image points.

In general, the HOG algorithm can be divided into 4 phases: gradient computation, orientation binning, descriptor blocks, and normalization blocks. In the first phase, the gradients are computed using Gaussian smoothing followed by discrete derivative masks. The experiments by Dalal and Triggs demonstrated that a simple 1-$D[-1, 0, 1]$ mask at none scale smoothing ($\sigma = 0$) works best. The second phase is responsible for dividing the gradient image into small-connected regions called cells (typical cell size is $6 \times 6$ or $8 \times 8$ pixels), and within each cell, a frequency histogram is computed representing the distribution of edge orientations within the cell. For this purpose, each pixel calculates a weighted function of the gradient magnitude (called vote, typically the magnitude itself is used) based on the orientation of the gradient element centered on it. Then, the edge orientations are quantized into $q$ bin uniformly spaced over 0–180° when an unsigned gradient is used, or 0–360° when a signed gradient is used. In a third phase, groups of adjacent cells are considered as spatial regions called blocks, (typical block size is $2 \times 2$ or $3 \times 3$ cells), and using an overlap between blocks significantly improves the algorithm performance. The grouping of cells into a block is the basis for the grouping and normalization of histograms. Two classes of block geometries commonly used are square or rectangular (R-HOG), with associations of spatial cells in squares or rectangles, and circular blocks (C-HOG) partitioned into cells in log-polar fashion. Finally, the blocks defined in the previous phase are normalized. For this purpose, let $v$ be the unnormalized block, $\|v\|_k$ be its $k$-norm for $k = 1{,}2$, and $\epsilon$ be a small normalization constant to avoid division by zero [41], then the following schemes can be used:

1. $L2 - norm\left(v \to v/\sqrt{\|v\|_2^2 + \epsilon^2}\right)$;

2. $L2 - Hys$ ($L2 - norm$ followed by clipping, limiting the maximum values of $v$ to 0.2, and renormalizing);

3. $L1 - norm\ (v \to v/\|v\|_1 + \epsilon)$;

4. $L1 - sqrt(v \to \sqrt{v/\|v\|_1 + \epsilon})$.

The final descriptor is then the vector of all components of the normalized cell responses from all of the blocks in the detection window.

On the other hand, it is common to find a support vector machine (SVM) classifier to complement a feature extractor HOG within a scheme of object recognition. One purpose of this paper is to determine the performance of an object recognition system that uses HOG and neural network-based classifier, particularly a multi-layer perceptron (MLP).

MLP was derived from the neuronal model known as perceptron, which was presented by Rossenblatt in 1958 [43]. The perceptron is based on the model of McCulloch and Pitts [44] and presents a learning rule based on error correction. MLP is an ANN composed of an input layer, $n$ hidden layers, and an output layer, all consisting of $m$ type-perceptron neurons. MLP is the most studied neural network model, which can approximate any continuous nonlinear function arbitrarily well on a compact interval. Due to this property, MLP became popular in order to parametrize nonlinear models and with classification purposes. Furthermore, the characteristics of the MLP are well known to solve the problem that occurs when the data available in training are generally not sufficient to cover the variability of the object's appearance. This problem is present in most recognition systems, including our own.

MLP belongs to the category of supervised classifiers. Then, with the training set, composed of $p$ pairs of input–output vectors that define the behavior of the system that the ANN will adapt, is defined as

$$\{(\mathbf{x}^1,\ \mathbf{t}^1),\ (\mathbf{x}^2,\ \mathbf{t}^2),\ \dots,\ (\mathbf{x}^p,\ \mathbf{t}^p)\} = \{(\mathbf{x}^\mu,\ \mathbf{t}^\mu)|\mu = 1,\ 2,\ \dots, p\} \tag{1}$$

where $\mathbf{x}^\mu = \left[x_i^\mu\right]_n$ and $\mathbf{t}^\mu = \left[t_k^\mu\right]_q$ are the input and target vectors, respectively.

MPL structure is defined as follows: It has an input layer of $n$ units, one or more successive hidden layers of intermediate units, and a layer of $q$ output units. Considering an MLP with $r$ hidden layers, where $\mathbf{x}^\mu = \left[x_i^\mu\right]_n$ is the $\mu$-th input vector that belongs to the training set (defined by **Eq. (1)**). Then, $(\mathbf{y}^l)^\mu = \left[\left(y_{j^l}^l\right)^\mu\right]_{m^l}$, $\mathbf{z}^\mu = \left[z_k^\mu\right]_q$ and $\mathbf{t}^\mu = \left[t_k^\mu\right]_q$ represent the output of the $l$-th hidden layer, the output generated by MLP and the output target that MLP must generate, respectively, when $\mathbf{x}^\mu$ is presented to the network; where $l = (1, 2, \dots, r)$, $m^l$ and $q$ indicate the number of units comprising the $l$-th hidden layer and the output layer, respectively.

Each unit's output of layer $l$ will be connected to the input of each unit in the layer $l + 1$, and a synaptic weight will be associated with each of these connections. Thus, the synaptic weights and thresholds of the first hidden layer are represented by $\mathbf{W}^1 = \left[w^1_{j^1 i}\right]_{m^1 \times n}$ and $\boldsymbol{\theta}^1 = \left[\theta^1_{j^1}\right]_{m^1}$, respectively. For remaining hidden layers, $l = (2, \ldots, r)$, the synaptic weights and thresholds are defined as $\mathbf{W}^l = \left[w^l_{j^l j^{l-1}}\right]_{m^l \times m^{l-1}}$ and $\boldsymbol{\theta}^l = \left[\theta^l_{j^l}\right]_{m^l}$. Finally, the synaptic weights and thresholds of the output layer are defined as $\mathbf{W}^{r+1} = \left[w^{r+1}_{kj^r}\right]_{q \times m^r}$ and $\boldsymbol{\theta}^{r+1} = \left[\theta^{r+1}_k\right]_q$.

MLP is a feed-forward neural network, and its operation is defined as follows. The output of first hidden layer is expressed mathematically as follows:

$$y^1_{j^1} = f\left(\sum_i w^1_{j^1 i} \bullet x_i - \theta^1_{j^1}\right) \tag{2}$$

Whereas the output of the $l$-th hidden layer, with $l = (2, \ldots, r)$, is computed by

$$y^l_{j^l} = f\left(\sum_{j^{l-1}} w^l_{j^l j^{l-1}} \bullet y^{l-1}_{j^{l-1}} - \theta^l_{j^l}\right) \tag{3}$$

Finally, the MLP operation is defined as:

$$z_k = g\left(\sum_{j^r} w^{r+1}_{kj^r} \bullet y^r_{j^r} - \theta^{r+1}_k\right) \tag{4}$$

Typically, activation functions for units of hidden layers, $f(\bullet)$, are nonlinear; e.g., unipolar sigmoid function $1/(1 + e^{-x})$ and bipolar sigmoid function $(1 - e^{-x})/(1 + e^{-x})$. Activation functions of this type introduce the nonlinearity into the network and enable the MLP to approximate any nonlinear function with arbitrary accuracy. The activation functions of the units in the output layer, $g(\bullet)$ may be linear or nonlinear, depending on the application.

The MLP evolution based its success on the design of training algorithms that could minimize the error committed by the network by adequately and automatically modifying the values of the synaptic weights. In this sense, the training algorithm called backpropagation is the most popular used to adapt the MLP to a specific application because it is conceptually simple and computationally efficient. In 1974 Paul Werbos developed the basic principles of backpropagation, while developing his PhD thesis, by implementing a system that estimated a dynamic model for predicting social communications and nationalism [45]. In 1986 Rumelhart et al. formalized the backpropagation algorithm as a method that allows a type-MLP ANN to learn the association that exists between a set of input patterns and the corresponding classes [46]. Over time, backpropagation has become one of the most widely used neural learning methods, proving to be an efficient tool in applications of pattern recognition, dynamic modeling, sensitivity analysis, and the control of systems over time, among others.

The backpropagation algorithm looks for the minimum error function in weight space using the method of gradient descent. This method is applied for training the units of hidden layers of an MLP; that is to say, the basic idea of this algorithm states that updating the synaptic weights of the units of a layer depends on the error generated by the layer itself and errors generated by the following layers. The aforementioned is established by the mathematical structure of the backpropagation algorithm, which can be expressed as follows:

$$\Delta w_{kj^r}^{r+1} = \varepsilon \bullet \delta z_k \bullet y_{j^r}^r, \tag{5a}$$

where $\delta z_k = (t_k - z_k) g'\left(u_k^{r+1}\right)$ and $u_k^{r+1} = \sum_{j^r} w_{kj^r}^{r+1} \bullet y_{j^r}^r - \theta_k^{r+1}$

$$\Delta w_{j^l j^{l-1}}^l = \varepsilon \bullet \delta y_{j^l}^l \bullet y_{j^{l-1}}^{l-1}, \tag{5b}$$

where $\delta y_{j^l}^l = f'^l_{j^l}\left(u_{j^l}^l\right) \sum_k \delta z_k \bullet w_{kj^l}^{l+1}$ and $u_{j^l}^l = \sum_{j^{l-1}} w_{j^l j^{l-1}}^l \bullet y_{j^{l-1}}^{l-1} - \theta_{j^l}^l$; for $l = (2, \ldots, r)$

$$\Delta w_{j^1 i}^1 = \varepsilon \bullet \delta y_{j^1}^1 \bullet x_i \tag{5c}$$

where $\delta y_{j^1}^1 = f'^1_{j^1}\left(u_{j^1}^1\right) \sum_{j^2} \delta y_{j^2}^2 \bullet w_{j^2 j^1}^2$ and $u_{j^1}^1 = \sum_i w_{j^1 i}^1 \bullet x_i - \theta_{j^1}^1 \varepsilon$ is a small-valued constant that defines the learning rate of the network.

## 3. Object recognition based on histogram of oriented gradients and multi-layer perceptron

The Object recognition system based on the histogram of oriented gradients and multi-layer perceptron (ORS HOG-MLP) proposed in this paper presents the following contributions: (1) offers good performance in multiclass applications. (2) Determine the performance of an object recognition system that uses HOG and neural network-based classifier, particularly an MLP. (3) In order to improve the characterization process of the image, a modification that improves the properties representation of the gradient calculation algorithm is proposed.

The ORS HOG-MLP is an algorithm that is geared to automatic object recognition. **Figure 1** shows the elements that are part of this system and the relationship that exists between them.

Initially, a window detector process takes samples over the entire image, each sample has a fixed size and it is called a detection window. Better performance of the algorithm occurs when there is an overlap between detection windows. The image will be sampled several times and on each occasion, the detection window size will be different. This feature is intended to avoid image segmentation and get an algorithm robust to the object's size.

Then the ORS HOG-MLP algorithm is applied to the extracted detection window. First, the gradient of an image at each pixel is computed. Let the discrete version of a detection window be represented as the matrix $\mathbf{A} = \left[a_{ij}\right]_{wi \times hi}$; $wi$ and $hi$ are the width and height of the detection window, respectively, and $a$ represents the $ij$-th pixel value, $0 \le a \le 2^n - 1$, and $n$ is the number of bits necessary to represent the value of a pixel. Now, the aim of this process is to compute the magnitude and direction of the gradients for each pixel. For this purpose, we use a 1-$D[-1, 0, 1]$ mask at none scale smoothing ($\sigma = 0$) which is applied over all image pixels. The magnitudes and directions obtained by this process are grouped into two matrices (**M** and **Th**); these matrices are defined as

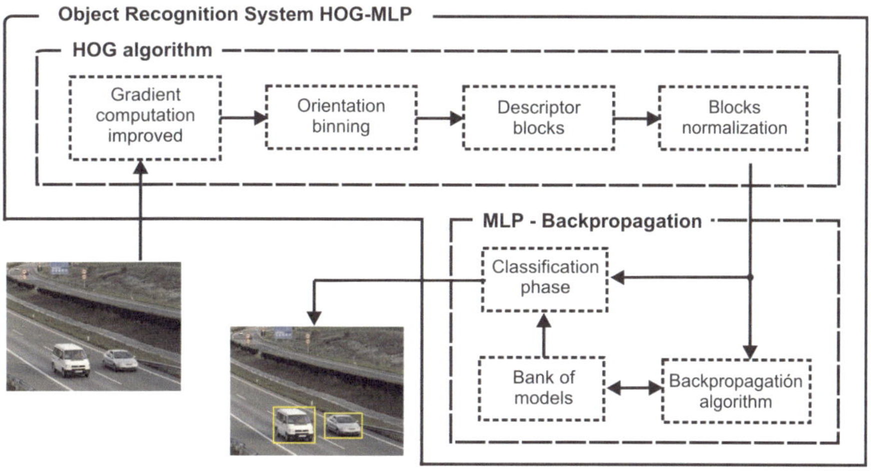


**Figure 1.**
*Block diagram of ORS HOG-MLP.*

$$\mathbf{M} = [m_{ij}]_{wi \times hi}; m_{ij} = \sqrt{[a_{i+1,j} - a_{i-1,j}]^2 + [a_{i,j+1} - a_{i,j-1}]^2}$$
$$\mathbf{Th} = [th_{ij}]_{wi \times hi}; th_{ij} = tan^{-1} \frac{a_{i,j+1} - a_{i,j-1}}{a_{i+1,j} - a_{i-1,j}} \tag{6}$$

**Figure 2** shows the results of gradient computation process. A detection window includes information that does not belong to the interest object (information about other objects and background information), during the gradient computation process, this information corrupts the interest object.

In order to improve the representation of the interest object properties by reducing this information, we propose to apply a threshold operation on the matrices **M** and **Th**,

$$m_{ij} = \begin{cases} 0, & \text{if } m_{ij} < umbral\_grad \\ n_{ij}, & \text{otherwise} \end{cases}$$
$$th_{ij} = \begin{cases} 0, & \text{if } m_{ij} < umbral\_grad \\ th_{ij}, & \text{otherwise} \end{cases} \tag{7}$$

**Figure 2** also shows the result obtained by applying this process on the matrices.

Now, both matrices **M** and **Th** are divided into $cx \times cy$ small-connected regions of $px \times py$ pixels, called magnitude cells and direction cells, and defined as $\mathbf{cm}_{lk} = \left[c\_m_{ij}^{lk}\right]_{px \times py}^{cx \times cy}$ and $\mathbf{cth}_{lk} = \left[c\_th_{ij}^{lk}\right]_{px \times py}^{cx \times cy}$; $cx = wi/px$ and $cy = hi/py$ (see **Figure 3**). Then, each pixel calculates a weighted function of its gradient magnitude based on its gradient orientation to contribute to the building of the histogram of the cell to which it belongs. For this purpose, a vector of $p = 9$ bins uniformly spaced over 0–180° is defined for the quantification process of gradient orientations: $\{bin_0, bin_1, \ldots, bin_8\} = \{10^\circ, 30^\circ, \ldots, 170^\circ\}$. The distance between bins is denoted by $d_{bins} = 20^\circ$.

Let $\mathbf{MC} = [\mathbf{Mc}_{lk}]$ be a matrix of $cx \times cy$ elements, called bins matrix, where the $\mathbf{Mc}_{lk} = \left[mc_o^{lk}\right]$ element is a $p$-dimensional vector containing the corresponding bins

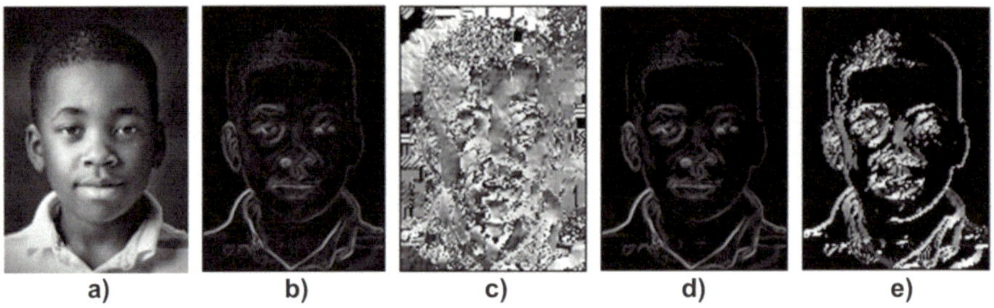


**Figure 2.**
*Results of gradient computation process. (a) Original image. (b) Gradient magnitude. (c) Pixel intensity proportional to the gradient direction (gradient direction). (d) Gradient magnitude improved. (e) Pixel intensity proportional to the gradient direction improved.*

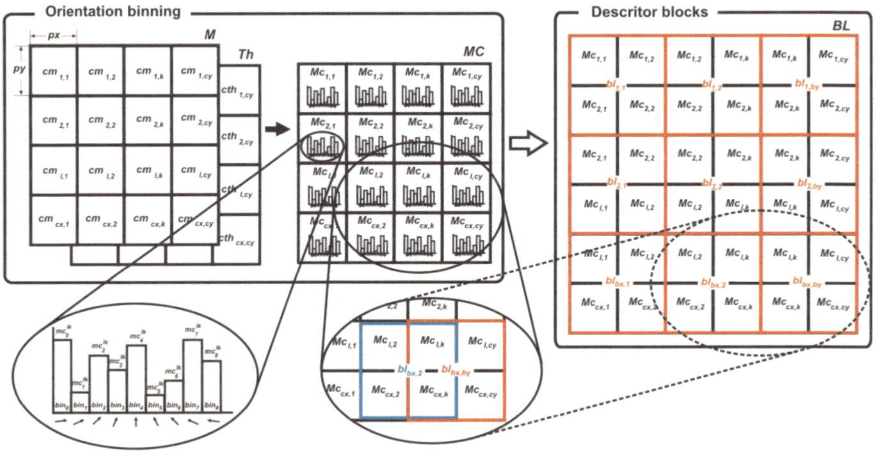


**Figure 3.**
*Details of orientation binning and descriptor block phases.*

$\left(mc_o^{lk}\right)$ of the histogram of the $lk$-th cell, $o = 0,1, \ldots ,p - 1$ (see **Figure 3**). Furthermore, considering two adjacent bins, $bin_o$ and $bin_{o+1}$, where $bin_o \leq c\_th_{ij}^{lk} \leq bin_{o+1}$, the distances between $c\_th_{ij}^{lk}$ and each of the bins are defined as $d_o = \left(c\_th_{ij}^{lk} - bin_o\right)$ and $d_{o+1} = \left(bin_{o+1} - c\_th_{ij}^{lk}\right)$.

Thus, the result of this process is defined as

$$
mc_o^{lk} = \begin{cases} mc_o^{lk} + \left(1 - \dfrac{d_o}{d_{bins}}\right) \bullet c\_m_{ij}^{lk}, & \text{if } bin_o < c\_th_{ij}^{lk} \\ mc_o^{lk} + c\_m_{ij}^{lk}, & \text{if } bin_o = c\_th_{ij}^{lk} \end{cases}
$$
$$
mc_{o+1}^{lk} = \begin{cases} mc_{o+1}^{lk} + \left(1 - \dfrac{d_{o+1}}{d_{bins}}\right) \bullet c\_m_{ij}^{lk}, & \text{if } c\_th_{ij}^{lk} < bin_{o+1} \\ mc_{o+1}^{lk} + c\_m_{ij}^{lk}, & \text{if } bin_{o+1} = c\_th_{ij}^{lk} \end{cases} \tag{8}
$$

**Figure 4** shows the result of applying orientation binning process.

In the next process, sets of $2 \times 2$ adjacent cells are grouped into spatial regions called descriptor blocks (n_**Mc** - number of **Mc** per block). In order to get better performance, the blocks are formed using an overlap of one cell on both $x$-axis and $y$-axis (see **Figures 3** and **4e**). The result of this process is the matrix $\mathbf{BL} = [\mathbf{bl}]$, composed of $bx \times by$ descriptor blocks; where $bx = 1,2, \dots, cx - 1$, $by = 1,2, \dots, cy - 1$ and a **bl** block is defined as

$$\mathbf{bl}_{bx,by} = \{\mathbf{Mc}_{bx,by}, \mathbf{Mc}_{bx,by+1}, \mathbf{Mc}_{bx+1,by}, \mathbf{Mc}_{bx+1,by+1}\} \tag{9}$$

Then, each descriptor block must be normalized. For this purpose, we decided to use the $L2 - Hys$ block normalization scheme, which first applied the $L2 - norm$ (scheme)

$$\mathbf{bl}_{bx,by} = \frac{\mathbf{bl}_{bx,by}}{\sqrt{\left\|\mathbf{bl}_{bx,by}\right\|_2^2 + \epsilon^2}} = \frac{\mathbf{bl}_{bx,by}}{\sqrt{\sum_o \left|mc_o^{bx,by}\right|^2 + \sum_o \left|mc_o^{bx,by+1}\right|^2 + \sum_o \left|mc_o^{bx+1,by}\right|^2 + \sum_o \left|mc_o^{bx+1,by+1}\right|^2 + \epsilon^2}} \tag{10}$$

Thus, each $mc.$ (bin) of the $bl_{bx,by}$ block is limited to a maximum value of 0.2: $mc = 0.2,|$ if $mc > 0.2$, and then it is renormalized again with $L2 - norm$.

Eventually, the final object descriptor, **Od,** is a $r$-dimensional vector of all components (bins) of the normalized cell responses from all of the blocks in the detection window

$$\begin{aligned}
\mathbf{Od} &= [od]_r = \{\mathbf{bl}_{1,1}, \mathbf{bl}_{1,2}, \dots, \mathbf{bl}_{bx,by}\} \\
&= \{\{\mathbf{Mc}_{1,1}, \mathbf{Mc}_{1,2}, \mathbf{Mc}_{2,1}, \mathbf{Mc}_{2,2}\}, \{\mathbf{Mc}_{1,2}, \mathbf{Mc}_{1,3}, \mathbf{Mc}_{2,2}, \mathbf{Mc}_{2,3}\}, \dots, \\
&\{\mathbf{Mc}_{bx,by}, \mathbf{Mc}_{bx,by+1}, \mathbf{Mc}_{bx+1,by}, \mathbf{Mc}_{bx+1,by+1}\}\} \\
&= \{\{\{mc_0^{11}, \dots, mc_8^{11}\}, \{mc_0^{12}, \dots, mc_8^{12}\}, \{mc_0^{21}, \dots, mc_8^{21}\}, \{mc_0^{22}, \dots, mc_8^{22}\}\}, \\
&\{\{mc_0^{12}, \dots, mc_8^{12}\}, \{mc_0^{13}, \dots, mc_8^{13}\}, \{mc_0^{22}, \dots, mc_8^{22}\}, \{mc_0^{23}, \dots, mc_8^{23}\}\}, \dots, \\
&\{\{mc_0^{bx,by}, \dots, mc_8^{bx,by}\}, \{mc_0^{bx,by+1}, \dots, mc_8^{bx,by+1}\}, \{mc_0^{bx+1,by}, \dots, mc_8^{bx+1,by}\}, \\
&\{mc_0^{bx+1,by+1}, \dots, mc_8^{bx+1,by+1}\}\}\}
\end{aligned} \tag{11}$$

where $r = bx \times by \times \text{n\_}\mathbf{Mc} \times p$ and $od_0 = mc_0^{11}, od_1 = mc_1^{11}, \dots, od_8 = mc_8^{11}, od_9 = mc_0^{12}, \dots, od_{17} = mc_8^{12}, \dots, od_{r-9} = mc_0^{bx+1,by+1}, od_{r-1} = mc_8^{bx+1,by+1}$

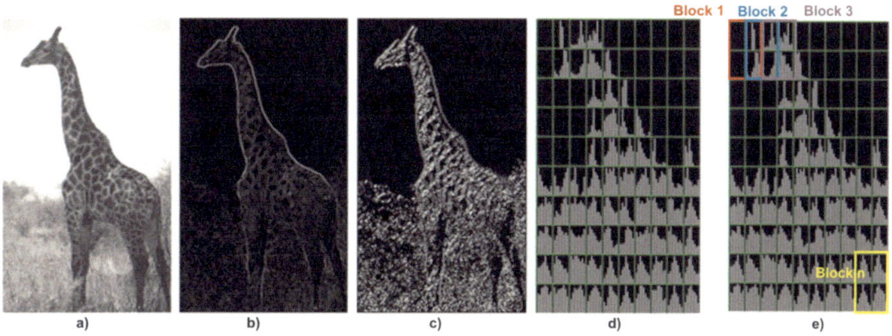


**Figure 4.**
*Result of orientation binning process. (a) Original image. (b) Gradient magnitude. (c) Gradient direction. (d) Histograms of the cells. (e) Descriptor blocks phase.*

**Figure 5** shows the sequence of processes that generate the HOG descriptor of the object.

Now, the MLP belonging to ORS HOG-MLP must be adapted to be able to recognize the interest objects. For this purpose, the processes described above are applied to $q$ interest objects and each descriptor obtained is associated with its corresponding class, $c = [c]_k$; thus, the training set of MLP is defined as

$$\{(\mathbf{Od}^1,\ \mathbf{c}^1),\ (\mathbf{Od}^2,\ \mathbf{c}^2),\ \ldots,\ (\mathbf{Od}^q,\ \mathbf{c}^q)\} = \{(\mathbf{Od}^\mu,\ \mathbf{c}^\mu)|\mu = 1, 2,\ \ldots, q\} \qquad (12)$$

The MPL structure is established as follows: It has one hidden layer of $h$ units, the input layer has $r$ units and the output layer has $v$ units. Considering the training set, the backpropagation learning algorithm is used to generate the bank of models, which includes the vital information of the objects that integrate the training set. Essentially, the bank of models is present in the synaptic weights, $\mathbf{W}^1 = \left[w^1_{jo}\right]_{h\times r}$ and $\mathbf{W}^2 = \left[w^2_{kj}\right]_{v\times h}$, that define the connections between neurons that integrate the MLP.

Finally, the operation process of ORS HOG-MLP, when the $\mathbf{Od}^\mu$ object is presented, is defined as

$$c^\mu_k = g\left(\sum_{j=0}^{h-1} w^2_{kj} \bullet f\left(\sum_{o=0}^{r-1} w^1_{jo} \bullet od^\mu_o - \theta^1_j\right) - \theta^2_k\right) \qquad (13)$$

where $k = 0,2, \ldots ,v$.

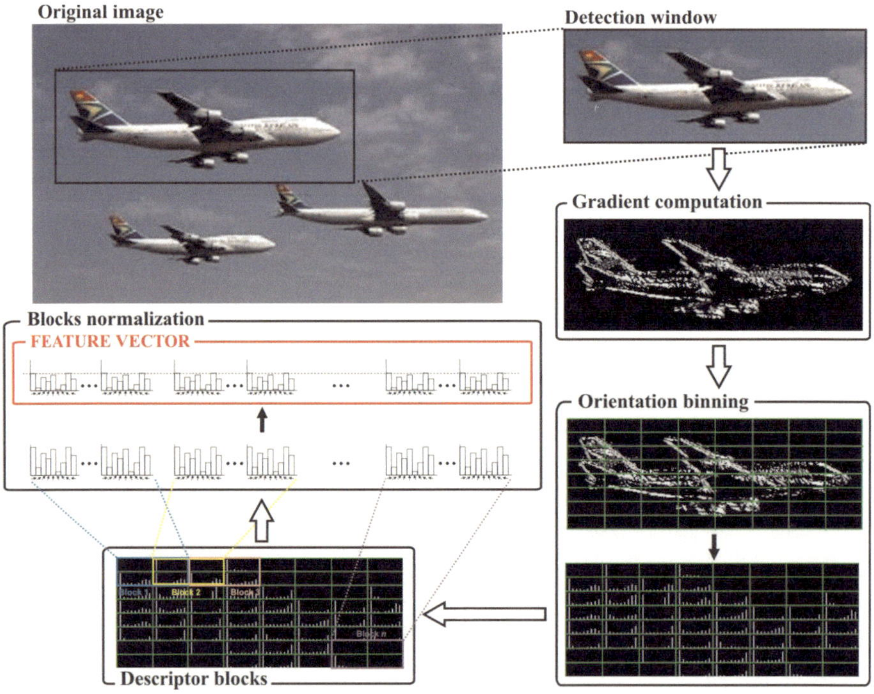


**Figure 5.**
*Sequence of processes for generating the HOG descriptor of the object.*

## 4. Experimental results and discussion

This section is intended to measure the ORS HOG-MLP performance through a set of experiments. In the first experiment, system performance is analyzed when it is adapted to recognize only one object class and was conducted on several object classes. In the second experiment, the system behavior is analyzed when it is configured to recognize multiple classes. Finally, our scheme's performance is with other results reported in the literature.

Before we present and discuss the data obtained from experiments, let us introduce concepts and methods used to measure ORS HOG-MLP performance. Let $O$ be an object presented to ORS HOG-MLP and the system generates the class $c^0$ to indicate that the object does not belong to its bank of models, then

A True Positive (TP) occurs when the class generated by the system is $\mathbf{c}^{\mu}$ and $O$ belongs to the class $\mathbf{c}^{\mu}$, for $\mu = 2, \ldots, q$; this indicates a successful classification.

A true negative (TN) occurs when the class generated by the system is $\mathbf{c}^0$ and $O$ does not belong to the bank of models; this indicates a successful rejection.

A False Positive (FP) occurs when the class generated by the system is $\mathbf{c}^{\mu}$ and $O$ does not belong to this class, for $\mu = 2, \ldots, q$; this indicates an incorrect classification.

A False Negative (FN) occurs when the class generated by the system is $\mathbf{c}^0$ and $O$ belongs to the bank of models; this indicates an incorrect rejection.

From these concepts, objective methods, which give information concerning system performance are obtained.

True positive rate (TPR). TPR determines the sensitivity of the system, i.e. it measures the proportion of successful classification obtained by system; TPR is defined as

$$\mathrm{TPR} = \frac{\text{number of TP}}{\text{number of TP} + \text{number of FN}} \tag{14}$$

False Positive Rate (FPR). FPR indicates the proportion of wrongly classified objects; FPR is defined as

$$\mathrm{E}(u) = (\mathrm{FPPW}(u), \mathrm{FNR}(u)) \tag{15}$$

Accuracy (ACC). ACC is used to evaluate the tendency of the system to ascertain correct TPs, defined as

$$\mathrm{ACC} = \frac{\text{number of TP} + \text{number of TN}}{\text{number of TP} + \text{number of TN} + \text{number of FP} + \text{number of FN}} \tag{16}$$

False negative rate (FNR). FNR indicates the miss rate of the system, defined as

$$\mathrm{FNR} = \frac{\text{number of FN}}{\text{number of TP} + \text{number of FN}} = 1 - \mathrm{TPR} \tag{17}$$

False positives per window (FPPW). FPPW indicates the number of errors by detection window and it is defined as

$$\mathrm{FPPW} = \frac{\text{number of FP}}{\mathrm{N}} \tag{18}$$

where N is the total number of windows processed. Finally, the proposed MLP classifier returns a real value for each detection window, this value is thresholding with a fixed value $u$ in order to determine whether or not it is an object belonging to a class of bank of models. Thus, FNR and FPPW are dependent functions of $u$, allowing the plotting of evaluation curves ROC (Receiver Operating Characteristic) [47], **Eq. (19)**, that show the tradeoff between the miss rate and the FPPW for each $u$.

$$E(u) = (FPPW(u), FNR(u)) \tag{19}$$

The Caltech 101 dataset, collected by Fei-Fei et al. [48], was used for benchmarking our proposal. For both training and operation phases of the system, positive images (those that contain only interest objects) and negative images (those that do not contain objects of interest or background) were generated.

The Caltech 101 dataset consists of a set of 1179 negative images for the system training phase $\left(I_{-}^{train}\right)$ and a set of 1179 negative images for the system operation phase $\left(I_{-}^{test}\right)$. The number of positive images for training and operation phases is $\left(W_{+}^{train\_\gamma}\right)$ and $\left(W_{+}^{test\_\gamma}\right)$, respectively, varies for each class (see **Table 2**).

In order to demonstrate the performance of our proposal when only one object needs to be identified, in first experiment, ORS HOG-MLP was adapted to individually identify each class of objects in **Table 2**. With the intention of adding robustness

| Data set | Positive images V-Scan | |
|---|---|---|
| | **Training** $\left(W_{+}^{train\_\gamma}\right)$ | **Testin** $\left(W_{+}^{test\_\gamma}\right)$ |
| Airplane | 200 $\left(W_{+}^{train\_cA}\right)$ | 200 $\left(W_{+}^{test\_cA}\right)$ |
| Butterfly | 40 $\left(W_{+}^{train\_cB}\right)$ | 40 $\left(W_{+}^{test\_cB}\right)$ |
| CarSide | 86 $\left(W_{+}^{train\_cC}\right)$ | 86 $\left(W_{+}^{test\_cC}\right)$ |
| Chair | 30 $\left(W_{+}^{train\_cD}\right)$ | 30 $\left(W_{+}^{test\_cD}\right)$ |
| Electric guitar | 30 $\left(W_{+}^{train\_cE}\right)$ | 30 $\left(W_{+}^{test\_cE}\right)$ |
| Faces | 100 $\left(W_{+}^{train\_cF}\right)$ | 100 $\left(W_{+}^{test\_cF}\right)$ |
| Helicopter | 40 $\left(W_{+}^{train\_cG}\right)$ | 40 $\left(W_{+}^{test\_cG}\right)$ |
| Horses | 85 $\left(W_{+}^{train\_cH}\right)$ | 85 $\left(W_{+}^{test\_cH}\right)$ |
| Ketch | 50 $\left(W_{+}^{train\_cI}\right)$ | 50 $\left(W_{+}^{test\_cI}\right)$ |
| Laptop | 40 $\left(W_{+}^{train\_cJ}\right)$ | 40 $\left(W_{+}^{test\_cJ}\right)$ |
| Motorbikes | 200 $\left(W_{+}^{train\_cK}\right)$ | 200 $\left(W_{+}^{test\_cK}\right)$ |
| Piano | 45 $\left(W_{+}^{train\_cL}\right)$ | 45 $\left(W_{+}^{test\_cL}\right)$ |
| Revolver | 40 $\left(W_{+}^{train\_cM}\right)$ | 40 $\left(W_{+}^{test\_cM}\right)$ |
| SoccerBall | 30 $\left(W_{+}^{train\_cN}\right)$ | 30 $\left(W_{+}^{test\_cN}\right)$ |

**Table 2.**
*Positive images per class.*

to the system, from $I_{-}^{train}$ and $I_{-}^{test}$, additional negative windows for training and testing of the system, $W_{-}^{train} = 1743$ and $W_{-}^{test} = 1909$, respectively, were generated. Thus, the training and testing of the final sets of negative images were defined as: $I_{-}^{\prime train} = I_{-}^{train} + W_{-}^{train} = 2922$, and $I_{-}^{\prime test} = I_{-}^{test} + W_{-}^{test} = 3288$.

Then, negative images are associated with class 0, and positive images of the object under study, defined by $\gamma$, to class 1. The training set is defined as $\left\{\left(\left\{I_{-}^{\prime train}\right\}, \boldsymbol{c}^{0}\right), \left(\left\{W_{+}^{train\_\gamma}\right\}, \boldsymbol{c}^{1}\right)\right\}$, this set is presented to ORS HOG-MLP in order to adapt it to recognize the object $\gamma$. On the other hand, the ORS HOG-MLP performance is evaluated by applying the recognition phase to sets $I_{-}^{\prime test}$ and $W_{+}^{test\_\gamma}$. This process was repeated for all objects belonging to **Table 2**.

The HOG algorithm parameters were adjusted as follows: The number of cells by detection window varies depending on the object shape, 9 bins uniformly spaced over 0–180° are defined, the size of blocks is of $2 \times 2$ adjacent cells and overlap of one cell in both $x$-axis and $y$-axis is used. The MLP is defined with the following features: The hidden layer has 5 neurons, the activation functions of neurons in the hidden layer and output layer are sigmoid, and random initial weights in the range of $[-0.25, 0.25]$, neurons bias is $-1.0$, learning rate $\varepsilon = 0.01$, and for each object, 20,000 iterations were carried out to train the network. **Figure 6** shows some examples of detections, and **Figure 7** and **Table 3** summarize the results of the first experiment.

Considering the results shown in **Table 3**, the average value of TPR and FPR parameters, 0.6387 and 0.001228, respectively, show that the probability of correctly classifying an object is approximately 64%, and the probability of misclassifying an object is less than 1%. Meanwhile, the ACC parameter indicates that the system accuracy is over 98% (e.g., for motorbikes, corresponding to 198 out

**Figure 6.**
*Examples of detection results on the Caltech 101 dataset. Detected objects are enclosed in rectangles.*

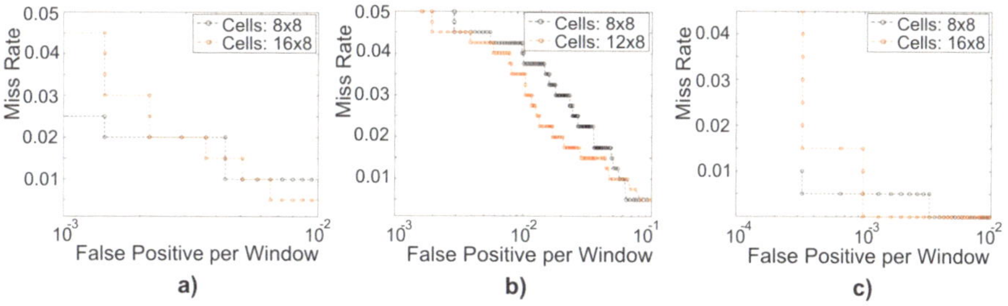


**Figure 7.**
*Performance of our proposal based on evaluation curves ROC: (a) Airplane class, (b) Butterfly class, (c) Motorbikes class.*

| Data set | Performance | | | | | Number of cells |
|---|---|---|---|---|---|---|
| | TPR | FPR | ACC | FP | FN | |
| Airplane | 0.955 | 0.00097 | 0.9963 | 3 | 9 | 16x8 |
| Airplane | 0.945 | 0.00060 | 0.9960 | 2 | 11 | 8x8 |
| Butterfly | 0.45 | 0.00130 | 0.9916 | 4 | 22 | 12x8 |
| Butterfly | 0.45 | 0.00190 | 0.9910 | 6 | 22 | 8x8 |
| Carside | 0.767 | 0.00097 | 0.9927 | 3 | 20 | 12x8 |
| Carside | 0.732 | 0.00097 | 0.9918 | 3 | 23 | 8x8 |
| Chair | 0.133 | 0.0 | 0.9916 | 0 | 26 | 12x8 |
| Chair | 0.233 | 0.00032 | 0.9923 | 1 | 23 | 8x8 |
| Electric guitar | 0.400 | 0.00032 | 0.9939 | 1 | 18 | 12x8 |
| Electric guitar | 0.400 | 0.00032 | 0.9939 | 1 | 18 | 8x8 |
| Faces | 0.960 | 0.00010 | 0.9987 | 0 | 4 | 16x8 |
| Faces | 0.760 | 0.00010 | 0.9924 | 0 | 24 | 8x8 |
| Helicopter | 0.375 | 0.00356 | 0.9884 | 11 | 25 | 16x8 |
| Helicopter | 0.500 | 0.00453 | 0.9891 | 14 | 20 | 8x8 |
| Horses | 0.741 | 0.0 | 0.9930 | 0 | 22 | 12x8 |
| Horses | 0.823 | 0.00388 | 0.9914 | 12 | 15 | 8x8 |
| Ketch | 0.680 | 0.00226 | 0.9926 | 7 | 16 | 10x10 |
| Ketch | 0.740 | 0.00064 | 0.9952 | 2 | 13 | 8x8 |
| Laptop | 0.650 | 0.00032 | 0.9952 | 1 | 14 | 16x16 |
| Laptop | 0.625 | 0.00064 | 0.9945 | 2 | 15 | 8x8 |
| Motorbikes | 0.960 | 0.00032 | 0.9972 | 1 | 8 | 16x8 |
| Motorbikes | 0.980 | 0.00001 | 0.9987 | 0 | 4 | 8x8 |
| Piano | 0.800 | 0.00064 | 0.9964 | 2 | 9 | 12x8 |
| Piano | 0.777 | 0.0 | 0.9968 | 0 | 10 | 8x8 |
| Revolver | 0.700 | 0.00194 | 0.9942 | 6 | 12 | 16x8 |
| Revolver | 0.675 | 0.00129 | 0.9945 | 4 | 13 | 8x8 |
| Soccerball | 0.241 | 0.00032 | 0.9926 | 1 | 22 | 12x12 |
| Soccerball | 0.241 | 0.00064 | 0.9923 | 1 | 22 | 8x8 |
| Watch | 0.660 | 0.00226 | 0.9871 | 7 | 34 | 12x8 |
| Watch | 0.700 | 0.00259 | 0.9880 | 8 | 30 | 8x8 |

**Table 3.**
*Results of first experiment.*

of 200 correct detections with 2 false positives). These results also show that using a detection window size of $8 \times 8$ or $16 \times 8$ cells does not significantly affect system performance.

In the definition of MLP structure, tests were performed using 5, 10, 15, and 20 neurons in the hidden layer. The results show variations of 0.5% of $10^{-3}$ FPPW, so we

DOI: http://dx.doi.org/10.5772/intechopen.106754

| Group | Objects number | Performance | | | | | | | | | | | | |
|---|---|---|---|---|---|---|---|---|---|---|---|---|---|---|
| | | TPR¹ | ACC¹ | TPR² | ACC² | TPR³ | ACC³ | TPR⁴ | ACC⁴ | TPR⁵ | ACC⁵ | TPR⁶ | ACC⁶ | #Cells |
| 1 | 6 | 0.025 | 0.976 | 0.03 | 0.979 | 1.000 | 0.989 | 0.660 | 0.983 | 0.75 | 0.985 | 0.103 | 0.980 | 8x8 |
| 2 | 6 | 0.930 | 0.988 | 0.670 | 0.983 | 0.100 | 0.983 | 0.970 | 0.991 | 0.600 | 0.987 | 0.630 | 0.981 | 8x8 |
| 2 | 6 | 0.960 | 0.993 | 0.770 | 0.989 | 0.030 | 0.986 | 0.995 | 0.996 | 0.680 | 0.992 | 0.630 | 0.984 | 12x8 |
| 3 | 6 | 0.960 | 0.993 | 0.180 | 0.985 | 0.130 | 0.987 | 0.970 | 0.99 | 0.450 | 0.988 | 0.975 | 0.994 | 8x8 |
| 4 | 4 | 0.935 | 0.994 | 0.710 | 0.990 | 0.350 | 0.990 | 0.955 | 0.995 | — | — | — | — | 8x8 |
| 4 | 4 | 0.945 | 0.993 | 0.670 | 0.988 | 0.28 | 0.987 | 0.975 | 0.995 | — | — | — | — | 12x8 |
| 5 | 3 | 0.920 | 0.993 | 0.700 | 0.989 | 0.950 | 0.995 | — | — | — | — | — | — | 8x8 |
| 5 | 3 | 0.930 | 0.994 | 0.690 | 0.990 | 0.960 | 0.996 | — | — | — | — | — | — | 16x8 |
| 6 | 2 | 0.950 | 0.995 | 0.980 | 0.997 | — | — | — | — | — | — | — | — | 8x8 |
| 6 | 2 | 0.975 | 0.997 | 0.980 | 0.997 | — | — | — | — | — | — | — | — | 12x8 |
| 7 | 2 | 0.930 | 0.997 | 0.600 | 0.994 | — | — | — | — | — | — | — | — | 8x8 |

*The groups are integrated as follows: group1 = {butterfly¹, hair², faces³, ketch⁴, laptop⁵, soccerball⁶}, group2 = {airplane¹, carside², electricguitar³, motorbikes⁴, revolver⁵, watch⁶}, group3 = {airplane¹, butterfly², chair³, faces⁴, laptop⁵, motorbikes⁶}, group4 = {airplane¹, carside², helicopter³, motorbikes⁴}, group5 = {airplane¹, carside², motorbikes³}, group6 = {airplane¹, motorbikes²}, group7 = {faces ¹, revolver²}.*

**Table 4.**
*Results of second experiment.*

decided to work with the smaller number of neurons to reduce the computational cost of the system.

In second experiment, ORS HOG-MLP is configured as a multi-class recognition system. Therefore, the system is trained to recognize several groups of objects belonging to **Table 2**, where the number of objects by group can be 2, 3, 4, or 6. Thus, the training set is defined as $\left\{\left(\left\{I'^{train}_{-}\right\}, c^0\right), \left(\left\{W^{train\_\gamma}_{+}\right\}, c^1\right), \ldots, \left(\left\{W^{train\_\gamma}_{+}\right\}, c^{\mu}\right)\right\}, \mu =$ 2,3,4,6, and $\gamma$ identifies the different objects belonging to a group; e.g. $\left\{\left(\left\{I'^{train}_{-}\right\}, c^0\right), (\{W^{train\_cA}_{+}\}, c^1), (\{W^{train\_cC}_{+}\}, c^2), (\{W^{train\_cK}_{+}\}, c^3)\right\}$ is a group that includes the airplane, carside, and motorbike objects, and the negative images are associated with the class 0. For this group, the recognition phase uses $I'^{test}_{-}$, $W^{test\_cA}_{+}$, $W^{test\_cC}_{+}$, and $W^{test\_cK}_{+}$ to evaluate system performance.

The configuration parameters of the HOG algorithm and MLP are the same as those used in Experiment 1. **Table 4** shows the results of this experiment. The results of second experiment show that system performance is not affected when operating in multiclass mode. This is deduced from the average values of TPR and ACC, which indicate that the probability of correctly classifying an object is approximately 68% and that the system accuracy fluctuates between 97% and 99%. e.g., for group2 = {airplane$^1$, carside$^2$, electricguitar$^3$, motorbikes$^4$, revolver$^5$, watch$^6$}, using a block of $8 \times 8$ cells, ORS HOG-MLP presents an ACC = 98% for airplane$^1$ (corresponding to 196 out of 200 correct detections with 4 false positives), and it presents an ACC = 99% for motorbikes$^4$ (corresponding to 198 out of 200 correct detections with 2 false positives). Those results make the robustness of the proposed system evident, when it is used in multi-object recognition applications.

Finally, the performance of the proposed scheme is compared with several object recognition schemes that have been cited frequently in related studies in this area. **Table 5** shows the results of this comparison.

**Table 5** shows a comparison of our scheme's performance with other results reported in the literature. With an ACC performance of 99%, our method presents a significant improvement over the previous result. The table also shows that the performance of the proposed scheme is not significantly affected when used in multi-object

| Method | | Data set | |
|---|---|---|---|
| | | Motorbikes | Cars |
| Object recognition scheme proposed in [35] | | 92.5% | 88.5% |
| Object recognition scheme proposed in [37] | | 99.0% | — |
| Object recognition scheme proposed in [38] | | 98.5% | 95.0% |
| Object recognition scheme proposed in [39] | | 97.4% | 96.7% |
| Object recognition scheme proposed in [40] | | 89.6% | 66.3% |
| HOG + MLP (our proposal, one object) | | 99.87% | 99.27% |
| HOG + MLP (our proposal, multi-objects) | 3 objects | 99.6% | 99.0% |
| | 4 objects | 99.5% | 99.0% |
| | 6 objects | 99.1% | 98.9% |

**Table 5.**
*Performance comparison on accuracy for object recognition for two of the Caltech categories with other methods from the literature.*

recognition applications. Zhang et al. reported a similar performance with 99% using a scheme composed of PCA-SIFT method and shape context method for object representation and two-layer AdaBoost network as classification technique [37]. However, due to the methods and techniques used by Zhang et al., this scheme presents a computational cost relatively greater than that presented by our proposal.

## 5. Conclusions

Although object recognition is a very active area of research, it is still considered an overly complex task due to the following difficulties: 1. Objects of the same class with high variability in appearance. A class of objects can integrate elements with variations in shape, color, and texture. In addition, multiple factors such as position, lighting, and occlusions, among others, can increase these differences. 2. The lack of reference images for the training phase of the classifier. The available data are generally not enough to cover the variability in appearance of objects. Furthermore, there may be significant differences in the conditions of training and system operation.

This research has placed special emphasis on the study of HOG algorithms for the feature extraction stage. This is because HOG has demonstrated that using normalized representations of objects can generate representations that provide discriminative information from the objects in an image. Furthermore, since HOG operates on local cells, it is invariant to geometric and photometric transformations as well as to changes in background and object position. On the other hand, it seeks to exploit the well-known features of the MLP to solve the problem that occurs when the limited data available during training are generally not enough to cover the variability in the appearance of objects.

It is important to emphasize that the proposed improvement in the step of calculating the HOG algorithm gradient reduces the rate of false positives. It was also demonstrated that HOG can accurately represent different objects and offers good performance in multiclass applications. Finally, we show that a classifier that uses a neuronal approach is an excellent complement to a HOG-based feature extractor.

It is the intention of this working group to use the proposed system in autonomous systems applications through its modeling on reconfigurable logic.

## Author details

Enrique Guzmán-Ramírez*, Ayax García, Esteban Guerrero-Ramírez, Antonio Orantes-Molina, Oscar Ramírez-Cárdenas and Ignacio Arroyo-Fernández
Technological University of Mixteca, Huajuapan de León, Oaxaca, México

*Address all correspondence to: eguzman@mixteco.utm.mx

## References

[1] Forsyth DA, Ponce J. Computer vision: A modern approach. 2dn ed. New Delhi: Pearson; 2011. p. 792

[2] Ballard DH, Brown CM. Computer Vision. New York: Prentice Hall; 1982. p. 539

[3] Mak KL, Peng P, Yiu KFC. Fabric defect detection using morphological filters. Image and Vision Computing. 2009;**27**(10):1585-1592. DOI: 10.1016/j.imavis.2009.03.007

[4] Abouelela A, Abbas HM, Eldeeb H, Wahdan AA, Nassar SM. Automated vision system for localizing structural defects in textile fabrics. Pattern Recognition Letters. 2005;**26**(10):1435-1443. DOI: 10.1016/j.patrec.2004.11.016

[5] Saeidi RG, Latifi M, Najar SS, Saeidi AG. Computer vision-aided fabric inspection system for on-circular knitting machine. Textile Research Journal. 2005;**75**(6):492-497. DOI: 10.1177/0040517505053874

[6] Li O, Wang M, Gu W. Computer vision based system for apple surface defect detection. Computers and Electronics in Agriculture. 2002;**36**(2–3, 223):215. DOI: 10.1016/S0168-1699(02)00093-5

[7] Kocer HE, Cevik KK. Artificial neural networks based vehicle license plate recognition. Procedia Computer Science. 2011;**3**:1033-1037. DOI: 10.1016/j.procs.2010.12.169

[8] Ozbay S. and Ercelebi E. Automatic vehicle identification by plate recognition. In Proceedings of World Academy of Science, Engineering and Technology; Turkey. 2005. pp. 222-225

[9] McKenna SJ, Jabri S, Duric Z, Rosenfeld A, Wechsler H. Tracking groups of people. Computer Vision and Image Understanding. 2000;**80**(1):42-56. DOI: 10.1006/cviu.2000.0870

[10] Comaniciu D, Ramesh V, Meer P. Kernel-based object tracking. Transactions on Pattern Analysis and Machine Intelligence. 2003;**25**(5):564-577. DOI: 10.1109/TPAMI.2003.1195991

[11] Wang X, Han TX, Yan S. An HOG-LBP human detector with partial occlusion handling. In: Proceedings of IEEE 12th International Conference on Computer Vision; 29 September – 2 October 2009; Japan. 2010. pp. 32-39. DOI: 10.1109/ICCV.2009.5459207

[12] Yang S, Liao X, Borasy U. A pedestrian detection method based on the HOG-LBP feature and gentle adaBoost. International Journal of Advancements in Computing Technology. 2012;**4**(19):553-560. DOI: 10.4156/IJACT.VOL4.ISSUE19.66

[13] Nandi CS, Tudu B, Koley C. An automated machine vision based system for fruit sorting and grading. In: Proceedings of Sixth International Conference on Sensing Technology; 18-21 December 2012; India. 2013. pp. 195-200. DOI: 10.1109/ICSensT.2012.6461669

[14] Viola P, Jones MJ. Robust real-time face detection. International Journal of Computer Vision. 2004;**57**(2):137-154. DOI: 10.1023/B:VISI.0000013087.49260.fb

[15] Muñoz-Salinas R, Aguirre E, García-Silvente M. People detection and tracking using stereo vision and color. Image and Vision Computing. 2007;**25**(6):995-1007. DOI: 10.1016/j.imavis.2006.07.012

[16] Zhao X, Lin Y, Ou B, Yang J. A wavelet-based image preprocessing

method or illumination insensitive face recognition. Journal of Information Science and Engineering. 2015;**31**(5): 1711-1731

[17] Yilmaz A, Javed O, Shah M. Object tracking: A survey. Journal ACM Computing Surveys. 2006;**38**(4):13. DOI: 10.1145/1177352.1177355

[18] Miura J, Kanda T, Shirai Y. An active vision system for real-time traffic sign recognition. In: Proceedings of Intelligent Transportation Systems; 1-3 October 2000; USA. 2002. pp. 52-57. DOI: 10.1109/ITSC.2000.881017

[19] Gonzales RC, Woods RE. Digital Image Processing. 4th ed. Ney York: Pearson; 2018. p. 1022

[20] Nixon M, Aguado A. Feature Extraction and Image Processing for Computer Vision. 3rd ed. London: Academic Press; 2012. p. 609

[21] Theodoridis S, Koutroumbas K. Patern recognition. 4th ed. London: Academic Press; 2009. p. 961

[22] Belongie S, Malik J, Puzicha J. Shape matching and object recognition using shape contexts. Transactions on Pattern Analysis and Machine Intelligence. 2002;**24**(24):509-522. DOI: 10.1109/34.993558

[23] Dalal N, Triggs B. Histograms of oriented gradients for human detection. In: Proceedings of Conference on Computer Vision and Pattern Recognition; 20-25 June 2005; USA. 2005. pp. 886-893. DOI: 10.1109/CVPR.2005.177

[24] Freeman WT, Roth M. Orientation histograms for hand gesture recognition. In: Proceedings of International Workshop on Automatic Face and Gesture Recognition. 1995. pp. 296-301

[25] Belongie S, Malik J, Puzicha J. Matching shapes. In Proceedings of 8th International Conference on Computer Vision; 7-14 July 2001; Canada. 2002. pp. 454-461. DOI: 10.1109/ICCV.2001.937552

[26] Lowe DG. Distinctive image features from scale-invariant keypoints. International Journal of Computer Vision. 2004;**60**(2):91-110. DOI: 10.1023/B:VISI.0000029664.99615.94

[27] Zhu Q, Avidan S, Yeh M, Cheng K. Fast human detection using a cascade of histogram oriented gradients. In: Proceedings of Conference of Computer Vision and Pattern Recognition; 17-22 June 2006; USA. 2006. pp. 1491-1498. DOI: 10.1109/CVPR.2006.119

[28] Kobayashi T, Hidaka A, Kurita T. Selection of histograms of oriented gradients features for pedestrian detection. In: Proceedings of Conference on Neural Information Processing; 13-16 November 2007; Japan. 2008. pp. 598-607. DOI: 10.1007/978-3-540-69162-4_62

[29] Socarras Y, Vázquez D, López AM, Gerónimo D, Gevers T. Improving HOG with image segmentation: Application to human detection. In: International Conference on Advanced Concepts for Intelligent Vision Systems; 4-7 September 2012; Czech Republic. 2012. pp. 178-189. DOI: 10.1007/978-3-642-33140-4_16

[30] Sermanet P, Eigen D, Zhang X, Mathieu M, Fergus R, LeCun Y. Overfeat: Integrated recognition, localization and detection using convolutional networks. In: International Conference on Learning Representation; 14-16 April 2014; Canada. 2014. pp. 1-16. DOI: 10.48550/arXiv.1312.6229

[31] He H, Chen S. Imorl: Incremental multiple-object recognition and

localization. Transactions on Neural Networks. 2008;**19**(10):1727-1738. DOI: 10.1109/TNN.2008.2001774

[32] Hanson SJ, Matsuka T, Haxby JV. Combinatorial codes in ventral temporal lobe for object recognition: Haxby (2001) revisited: Is there a "face" area? NeuroImage. 2004;**23**(1):156-166. DOI: 10.1016/j.neuroimage.2004.05.020

[33] Markou M, Singh S. Novelty detection, a review-part2: Neural network based approach. Signal Processing. 2003;**83**(12):2499-2521. DOI: 10.1016/j.sigpro.2003.07.019

[34] Guoqiang PZ. Neural networks for classification: A survey. Transactions on Systems, Man, and Cybernetics – Part C. Applications and Reviews. 2000;**30**(4): 451-462. DOI: 10.1109/5326.897072

[35] Fergus R, Perona P, Zisserman A. Object class recognition by unsupervised scale-invariant learning. In: Proceedings of Conference on Computer Vision and Pattern Recognition; 18-20 June 2003; USA. 2003. pp. II-II. DOI: 10.1109/ CVPR.2003.1211479

[36] Dempster AP, Laird NM, Rubin DB. Maximum likelihood from incomplete data via the EM algorithm. Journal of the Royal Statistical Society. Series B (Methodological). 1977;**39**(1):1-38

[37] Zhang W, Yu B, Zelinsky GJ, Samaras D. Object class recognition using multiple layer boosting with heterogeneous features. In: Proceedings of Conference on Computer Vision and Pattern Recognition; 20-25 June 2005; USA. 2005. pp. 323-330. DOI: 10.1109/ CVPR.2005.251

[38] Zhang J, Marszałek M, Lazebnik S, Schmid C. Local features and kernels for classification of texture and object categories: A comprehensive study. International Journal of Computer Vision. 2007;**73**(2):213-238. DOI: 10.1007/s11263-006-9794-4

[39] Leibe B, Leonardis A, Schiele B. Robust object detection with interleaved categorization and segmentation. International Journal of Computer Vision. 2008;**77**:259-289. DOI: 10.1007/ s11263-007-0095-3

[40] Laptev I. Improving object detection with boosted histograms. Image and Vision Computing. 2009;**27**(5):535-544. DOI: 10.1016/j.imavis.2008.08.010

[41] Dalal N. Finding People in Images and Videos [Doctoral thesis]. Saint Ismier, Frace, Institut National Polytechnique de Grenoble-INPG. 2006

[42] Kim S, Cho K. Design of high-performance HOG feature calculation circuit for real-time pedestrian detection. Journal of Information Science and Engineering. 2015;**31**(6):2055-2073. DOI: 10.6688/JISE.2015.31.6.13

[43] Rosenblatt F. The perceptron: A probabilistic model for information storage and retrieval in the brain. Psychological Review. 1958;**65**:386-408

[44] McCulloch WS, Pitts W. A logical calculus of the ideas immanent in nervous activity. The Bulletin of Mathematical Biophysics. 1943;**5**(4): 115-133

[45] Werbos PJ. Beyond regression: New tools for prediction and analysis in the behavioral sciences [Doctoral thesis]. Cambridge, MA, Committee on Appl. Math., Harvard Univ.; 1974

[46] Rumelhart DE, Hinton GE, Williams RJ. Learning Internal Representations by Error Propagation. San Diego: Univ. of California, Inst for Cognitive Science; 1985

[47] Fawcett T. An introduction to ROC analysis. Pattern Recognition Letters. 2006;**27**:861-874. DOI: 10.1016/j.patrec.2005.10.010

[48] Fei-Fei R, Fergus R, Perona P. Learning generative visual models from few training examples: An incremental Bayesian approach tested on 101 object categories. Computer Vision and Image Understanding. 2007;**106**(1):59-70. DOI: 10.1016/j.cviu.2005.09.012

Chapter 3

# EEG and MRI Processing for Alzheimer's Diseases

*Elias Mazrooei Rad*

## Abstract

A new method for the diagnosis of Alzheimer's disease in the mild stage is presented according to combining the characteristics of electroencephalogram (EEG) signal and magnetic resonance imaging (MRI) images. Then, proper features of brain signals are extracted according to the nonlinear and chaotic nature of the brain such as Lyapunov exponent, correlation dimension, and entropy. These features combined with brain MRI images properties include medial temporal lobe atrophy (MTA), cerebrospinal fluid flow (CSF), gray matter (GM), index asymmetry (IA), and white matter (WM) to diagnose the disease. Then two classifiers, the support vector machine and Elman neural network, are used with the optimal combined features extracted by analysis of variance. Results showed that between the three brain signals, and between the four modes of evaluation, the accuracy of the Pz channel and excitation mode was more than the others The accuracy of the results in Elman neural network with the combination of brain signal features and medical images is 94.4% and in the case without combining the signal and image features, the accuracy of the results is 92.2%.

**Keywords:** EEG, MRI, Alzheimer's diseases, SVM, Elman neural network

## 1. Introduction

Alzheimer's disease is a progressive disease of the mental faculties commonly seen in the elderly. Significant symptoms of this disease are memory loss, judgment, and important behavioral changes in the person [1]. The disease results in the loss of synapses of neurons in some areas of the brain, necrosis of brain cells in different areas of the nervous system, the formation of spherical protein structures called aging plaques outside neurons in some areas of the brain, and fibrous protein structures called neurofibrillary Tangles. A spiral is identified in the cell body of neurons. There is currently no definitive diagnosis or treatment for this disease. The prevalence of Alzheimer's disease is increasing rapidly [2]. The number of Alzheimer's patients in Iran has almost doubled in 13 years, according to the Iranian Alzheimer's Association. On the other hand, the costs of treatment, as well as care and nursing of these patients, are very high and difficult. This disease causes various mental disorders in the patient. It usually takes several years from the first signs of the disease to the acute stages of the disease, when most of the brain cells are destroyed. If this disease is not detected in time, new and up-to-date treatment methods will not work. The solution is to accurately identify the mechanism of this disease and its effect on brain signals, which is very difficult due to the dynamic nature of EEG signals and medical images, which due to the complex nature of this disease as a result, we must determine the best and most effective indicator to identify this disease and how this indicator relates to the characteristics of the brain signal and medical images. Medical image analysis has become very important in the diagnosis of mild Alzheimer's disease in recent years [3]. The high volume and complexity of medical images make early detection of Alzheimer's disease difficult for physicians and increase the workload of radiologists, in which case the use of computer-aided diagnostic (CAD), including image processing technologies, can help to increase the accuracy of diagnosis. The use of machine learning systems and deep processing of medical images with proper labeling and feature extraction can be one of the effective methods of diagnosing this disease [4]. Deep learning methods and machine learning techniques can be two effective and accurate methods in the early diagnosis of Alzheimer's disease [5]. Hippocampal volume analysis is used in medical image processing to diagnose mild Alzheimer's disease. Because before the atrophy creation, analyzing the volume of hippocampal material in MRI images can be used with deep processing techniques to extract proper features to identify mild-Alzheimer's disease [6]. 3D segmentation of MRI images further helps researchers diagnose Alzheimer's disease and obtain important information [7]. Determining the degree of atrophy of MRI images is an effective method for early detection of Alzheimer's disease. Also, assessing the degree of asymmetry in both the right and left hemispheres and analyzing volumetric mismatch can differentiate from mild to severe Alzheimer's disease [7]. Using statistical features of signal and obtaining temporal information and using spatial features of MRI images is an effective method for the more accurate evaluation of Alzheimer's disease [8]. Cortical atrophy means the gradual destruction of the nerve cells that make up the upper regions of the brain, specifically the structures found in the cerebral cortex, mostly due to a reduction or loss of oxygen and nutrients in these areas. There are also different methods for evaluating the medial temporal lobe that has different functional accuracy [9]. Longitudinal T1-weighted MRI studies are another effective way to distinguish mild-Alzheimer's patients from healthy ones [10]. Also, extracting the appropriate characteristics and deciding on the classification in this field are among the issues to be considered. Currently, there are several methods to diagnose this disease, and it is important to examine two issues in these methods. The cost of these methods and the acceptable accuracy are the points under consideration, so it seems necessary to identify a low-cost method with appropriate accuracy and precision. Therefore, in addition to extracting proper features of EEG signals and MRI images for AD diagnosis, another aim of this study is to identify the proper multimodal combining of these extracted features to increase the accuracy of mild-Alzheimer's disease detection by using a proper classifier.

## 2. Materials and methods

In this study, 40 volunteers were used to record brain signals and MRI images in healthy, mild, and severely ill groups. The number of subjects is 19 in the healthy group, 11 in the mild patient group, and 10 in the severely ill group. Forty volunteers with an age range of 60 to 88 years have been used to record brain signals. All participants in all groups were right-handed. Nineteen participants in the Mini-Mental State Exam (MMSE) test scored between 23 and 30 and were included in the group of healthy people. Eleven participants with an MMSE score of 19 to 22 were classified as mild-Alzheimer's patients, and finally, 10 participants with an MMSE score between 3 and 18 were included in the group of severe Alzheimer's patients.

The Powerlab SP device with two amplifiers was used to record the brain signal. In this device, three channels for recording the brain signal and one channel for recording the EOG signal, and the other channel for the external audio signal for stimulation have been used so that the stimulation signal and ERP do not occur simultaneously. The signal was recorded in 4-channel mode according to the standard 10–20. The sampling rate of the device is 1 kHz and 16 bits for each sample. Recording the brain signal in the form of subject training, recording the closed eye for 1 minute, recording the open eye for 1 minute, and recording while performing the task assigned to the subject, which includes A. Remembering the displayed shapes; B. The counting of target and non-target sounds in an oddball auditory test. After proper labeling by the physician in segregation of healthy individuals, and mild and severe Alzheimer's patients by MMSE test in the first part, how to record the brain signal in four steps is explained to people and they are asked to relax during Keep records to prevent the formation of motion artifacts and other unwanted factors, and this method of registration does not cause harm to the person. After preparing the subject, we perform the second step. In the closed eye mode, we record the signal for 1 minute. Then in the third step, the subject will be asked to open the eyes and record the signal for 1 minute. At the end of the third stage, the displayed images have no color so that the color feature does not have a different effect on different subjects. These images are displayed for 1 minute and after this time, the subject will be asked to close the eyes and recall the images (review the images) in the mind. Meanwhile, brain signals will be recorded for 1 minute. Participants will then be asked to open their eyes and express the shapes one by one aloud. In the last part, a sound with a frequency of 1 kHz called non-target sound and 1.5 kHz sound called target sound will be given to the subject. Before playing these two categories of sound, this step has been taught to the subject. In the fourth part of section (b), these sounds are played, and the subject is asked to press the right key as soon as hearing the target sound and the left key as soon as hearing the non-target sound. The interval between stimuli (sound playback) is 2 seconds and the sounds will be played randomly. The only important point is that 75% of the number of stimuli is non-target sound and 25% of the number of stimuli is target sound. If we assume the total number of stimuli to be 120, we will have a total of 30 target stimuli and 90 non-target stimuli, which can be randomly distributed between the non-target stimuli at 2-second intervals. The playing time of each sound (target and non-target) will be 300 milliseconds. This recording section is 276 seconds with the assumption of 120 excitations and the total signal recording time for each subject is approximately 10 minutes. A sample image of recording brain signals is shown in **Figure 1**.

The first step in processing brain signals is to eliminate high- and low-frequency noise and interference and to remove motion artifacts. It is clear that the removal of unwanted factors such as motion artifacts, signal deviation from the baseline, high- and low-frequency noise, and reduction of sampling rate is necessary for proper processing of brain signals and extraction of optimal features, and this increases the accuracy of brain signal processing [11]. The motion artifacts in the brain signal are caused by contractions of the muscles of the head and neck as well as the movement of the electrode. On the other hand, transpiration also causes frequency interference. To eliminate these artifacts and noise from the city, a pass filter with cutoff frequencies of 0.5 to 45 Hz was used [12].

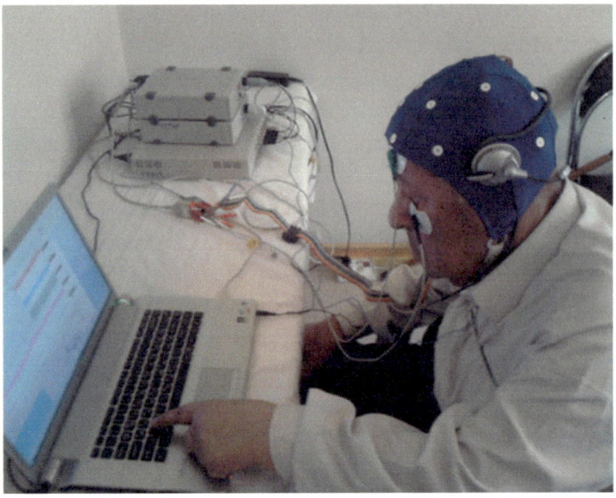

**Figure 1.**
*A sample research participant during brain signal recording.*

Neuroimaging techniques, physiological signs, and genetic analysis are methods used to diagnose Alzheimer's disease [13]. To detect Alzheimer's disease in its early stages, neuroimaging methods are used, which include SPECT, PET, and magnetic resonance imaging. The problem with SPECT and PET is the risks of radiation and its very high cost, time-consuming, and inconvenient. Therefore, apart from all these neuroimaging methods, MRI imaging is one of the standard methods used to diagnose Alzheimer's disease. The advantage of this method is the ease of registration and economic cost over the above methods. MRI images should be at least 3 Tesla and the slices should be 3 mm thick so that acceptable images can be seen to examine the lesions of aging coils and spiral plaques. The MRI image is displayed in three different directions in **Figure 2**. Then, the appropriate image segmentation, mask, and sharp filter are used for pre-processing.

Various diagnostic tools from the clinical and processing areas for early diagnosis of Alzheimer's disease have been reviewed. Methods of blood tests, speech therapy, physical function, and hearing status were first examined by a physician and then diagnosed with mild Alzheimer's disease by recording an electroencephalogram and combining it with medical images [14]. First, the EEG signal is recorded from three channels Fz, Cz, Pz as unipolar and then MRI images from the peritoneal area. Combining MRI images and EEG signals can be a way to diagnose mild Alzheimer's disease. Medial temporal lobe atrophy, cerebrospinal fluid, white and gray matter volume, and asymmetry between the two hemispheres are effective features in MRI images to diagnose Alzheimer's disease. Another approach to EEG signal analysis is nonlinear and dynamic signal methods. The parameters that express nonlinear behavior are dual. The first category is parameters that emphasize the dynamics of signal behaviors such as entropy and Lyapunov's exponent. These parameters describe how the system behaves over time. The second category emphasizes the geometric nature of motion paths in state space, such as the correlation dimension. In this view, the system is allowed to move in the adsorption bed at the appropriate time and then, the geometric dimension of the adsorption bed is obtained. One of the most important tools used to understand the behavior and dynamics of time series of vital signals, which are mainly extracted from nonlinear systems, is the phase diagram.

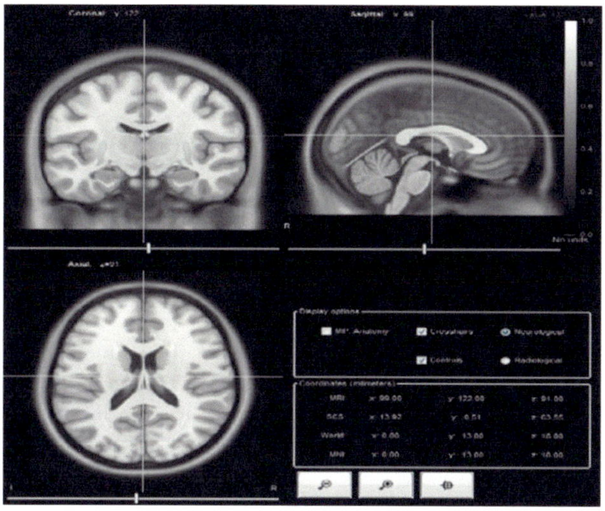

**Figure 2.**
*MRI image in three different modes.*

Using this tool, the behavioral characteristics and chaotic nature of the data can be demonstrated appropriately and qualitatively, as well as important parameters such as the path of the system in the state space [15]. In order to draw this diagram using the recorded time series, it is enough to draw each sample at any time in terms of another sample in the previous time. **Figure 3** shows the two-dimensional phase curve, and **Figure 4** shows the three-dimensional phase curve of the Fz, Cz, and Pz channels of the EEG signal from a healthy person with the eyes closed.

Lyapanov's exponent shows the average convergence or divergence of the trajectory path in the phase space. The correlation dimension shows the number of

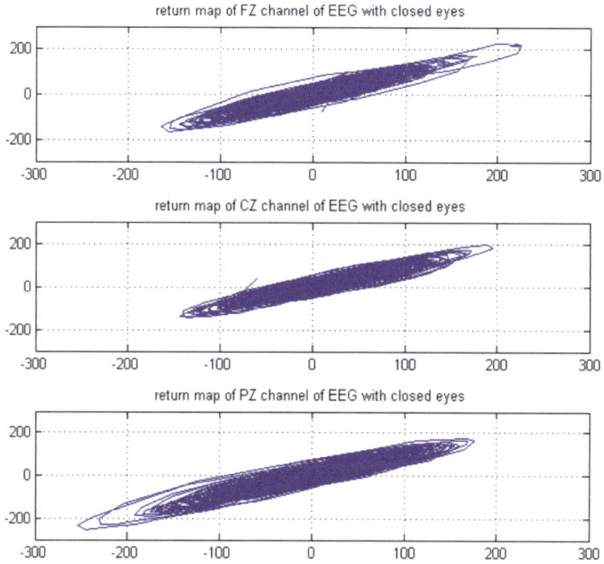


**Figure 3.**
*Two-dimensional phase curve of Fz, Cz, Pz channels EEG signal in closed eye.*

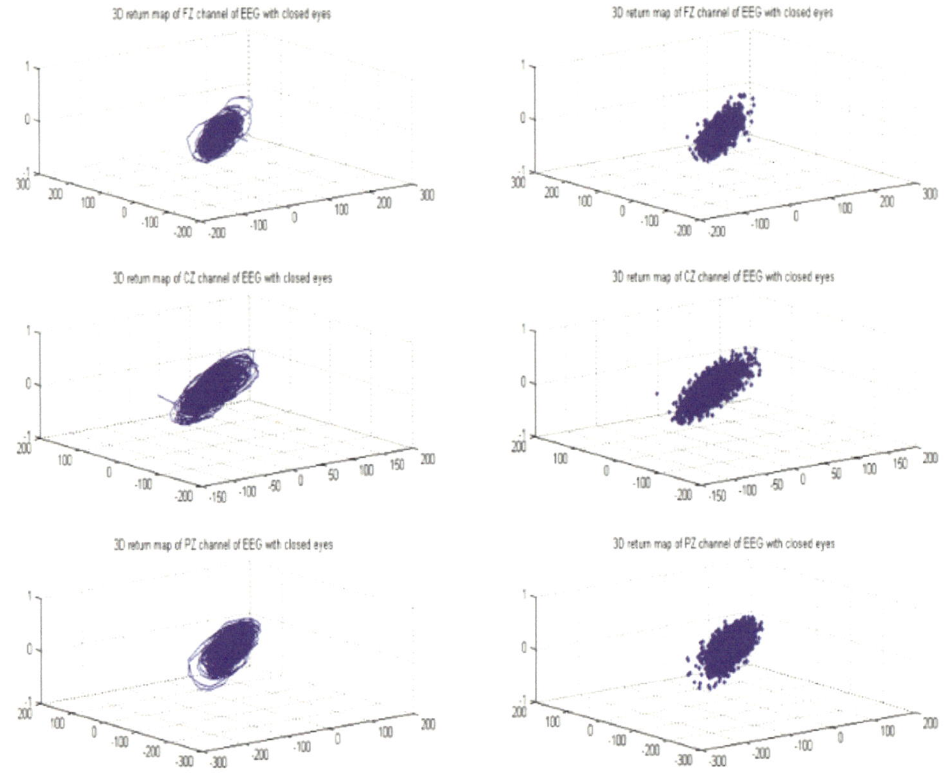


**Figure 4.**
*Three-dimensional phase curve of Fz, Cz, Pz channels EEG signal in closed eye.*

independent variables needed to describe the dynamics of the system and is another way to examine the chaotic signal. If the correlation dimension of a path is zero, it represents a steady state of the system, and if the value is equal to one, it represents prodigal behavior. The value of this variable is incorrect when chaotic behavior occurs. The higher the value of this parameter, the more complex the nonlinear system. Therefore, it can be said that the correlation dimension is the degree of complexity of the distribution of points in the phase space. **Figure 5** compares the correlation dimension of Pz channels between 3 groups of healthy people, mild patients, and severe patients. The amount of this feature decreases with the severity of the disease, which is evident in the Pz channel.

In patients with a clinical diagnosis of Alzheimer's disease, atrophy of the inner part of the temporal lobe is evident [16]. In the autosomal dominant form of Alzheimer's disease, atrophy of the inner part of the temporal lobe in patients, compared with controls, can be detected up to 3 years before the onset of clinical signs of cognitive impairment. In patients with Alzheimer's disease, hippocampal atrophy was reduced (10–50%), the amygdala was reduced to 40%, and parahippocampus was reduced to 40% compared with the control group, which was standardized for age. There is compelling evidence that atrophy of the internal structures of the temporal lobe, especially the hippocampus and entorhinal cortex, occurs early in the course of the disease and even before the onset of clinical symptoms [17]. The severity of changes in imaging of healthy elderly people makes it difficult to use MRI

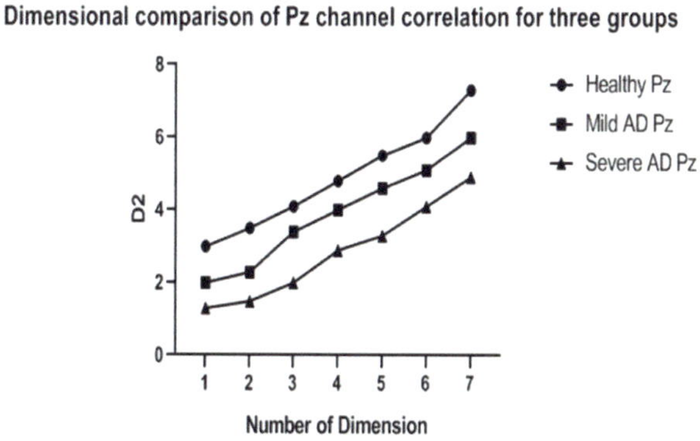


**Figure 5.**
*Dimensional comparison of Pz channel correlation for three groups of healthy subjects, mild patient and severe patient.*

as a definitive diagnostic method. By the time mild symptoms appear, the volume of the hippocampus may have decreased by more than 25%. Clinically, a reduction in hippocampal volume is associated with the severity of clinical signs and symptoms of memory loss, the patient's score on cognitive evaluation tests, and pathological findings. **Figure 6** shows the determination of spinal atrophy and asymmetry. However, another group believes that there is no clear association between lesions in the course of dementia, including lesions of hyperexcitability of white matter on MRI, and the severity of the symptoms of post-adjustment cognitive impairment for age. They believe that due to the high sensitivity of MRI in the diagnosis of hyperexcitability lesions in T2 view and on the other hand the low specificity of these lesions in the diagnosis of the disease, there is a weak relationship between MRI findings and clinical and neuropathological symptoms. Eq. (1) shows how to determine medial temporal lobe atrophy (MTA):

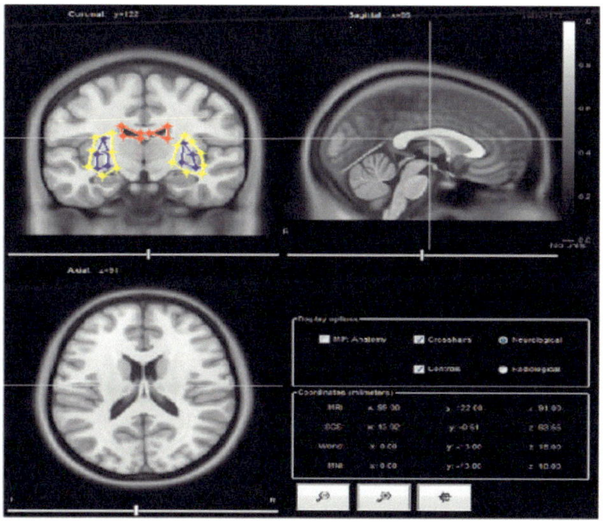

**Figure 6.**
*How to determine atrophy in MTA images.*

A: Internal temporal loop area.
B: Hippocampus and parahippocampus.
C: Unilateral lateral ventricle.

$$\text{MTAi} = (\text{A} - \text{B}) \times 10 / \text{C} \tag{1}$$

$$\text{Cleft}: \text{Area} = 187.3\,\text{mm2}, Avg = 691.1, Dev = 128.3.$$

$$\text{Cright}: \text{Area} = 173.1\,\text{mm2}, Avg = 648.2, Dev = 146.2.$$

$$\text{Aleft}: \text{Area} = 324.7\,\text{mm2}, Avg = 323.9, Dev = 238.1.$$

$$\text{Aright}: \text{Area} = 325.5\,\text{mm2}, Avg = 245.3, Dev = 191.3.$$

$$\text{Bleft}: \text{Area} = 200.3\,\text{mm2}, Avg = 190.8, Dev = 121.6.$$

$$\text{Bright}: \text{Area} = 220.1\,\text{mm2}, Avg = 160.4, Dev = 118.3.$$

When MTAi is calculated, two values are determined that each corresponds to a hemisphere. According to MTAi, the asymmetry index is calculated as Eq. (2), and the mean values of MTAi and IA for the three groups are given in **Table 1 (Figure 7)**.

$$IA = (\text{lMTAi} - \text{dMTAi}) / (\text{lMTAi} + \text{dMTAi}) \times 100 \tag{2}$$

Measurement of cerebrospinal fluid, gray matter, and white matter volumes from MRI images has been used to diagnose mild Alzheimer's disease [18].

In this study, nonlinear property that reflects the dynamic nature of the brain signal, including Lyapunov exponent and correlation dimension, is also determined. On the other hand, in order to determine the optimal characteristics in three classes of healthy people, mild patients, and severe patients, the method of analysis of variance has been used. Brain signals from the three channels Fz, Cz, and Pz are recorded in four modes: closed-eye, open-eye, reminder, and stimulus. Forty-five properties are specified in the excitation mode. **Table 2** shows the results of analysis of variance for three channels of Fz, Cz, and Pz between the group of healthy individuals, and mild and severe patients [19]. This analysis method is used in the classification of three or more classes to determine the optimal and effective characteristics.

| Group | Mean MTAi | Mean IA |
|---|---|---|
| Healthy | 2.3 | 1.7 |
| Mild AD | 4.5 | 2.5 |
| Severe AD | 5.7 | 3.3 |

**Table 1.**
*Mean values of MTAi and IA for the three groups are given.*

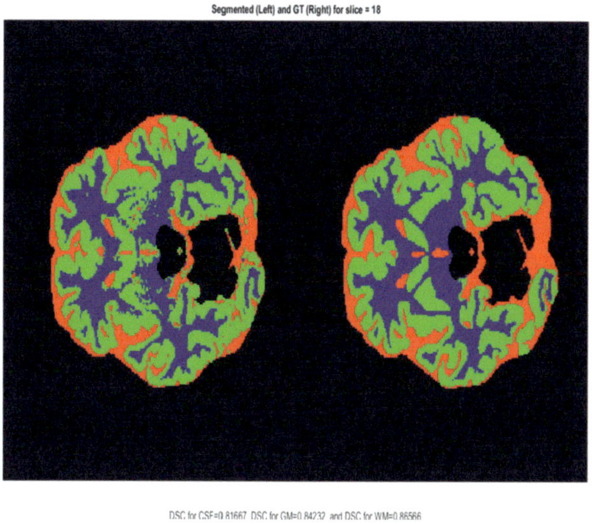


**Figure 7.**
*Cerebrospinal fluid, gray and white matter volume in the image in 18th slice of a participant with mild AD.*

| Channel | Benefit features with ANOVA | Number of benefit feature |
|---|---|---|
| Fz | A-band power,D3-absolute Mean,D3-average power,D3-Std,Coherence | 5 |
| Cz | A-band power,D3-absolute Mean,D3-average power,D3-Std,Colierence,L-ave,D2 | 7 |
| Pz | T-band power, A-band power,D3-absolute Mean,D3-average power,D3-Std, Coherence,L-ave,D2,ApEn | 9 |

**Table 2.**
*Compare the number and types of optimum features between Fz, Cz, Pz channels.*

In this study, the purpose of using the classifier, after extracting the optimal characteristics of brain signals, was to separate the three groups of healthy people, mild and severe patients, and two classifiers such as SVM and the Elman neural network been used, aiming at comparing static and dynamic classifiers [20]. One of the classification methods with the teacher is the backup vector machine method. This view is based on statistical learning theory, but its implementation is similar to the neural network. This method was designed to separate data into two categories. Of course, if you use several SVMs in parallel and with different methods, this method can be used to classify data into more than two categories. SVM claims: It solves the major problem of neural networks, namely overfitting [21]. The results of EEG signal accuracy of different channels in different modes are determined at two levels (mild-severe, mild-healthy, and healthy-severe). Due to the linear separator for three classes containing poor results accuracy, levels are divided into two levels. In this study, a 2-layer Elman neural network was used which has 8 neurons in the latent layer and 1 neuron in the output layer [22]. The number of neural network inputs is equal to the number of features, and the number of hidden layer neurons is equal to the number of optimal features, and to determine the best results, various experiments with different numbers of neurons in the hidden layer have been performed. In the hidden layer and the output, the Sigmoid activation function is used due to its nonlinear property. There are many training functions for teaching the Elman network, wherein in this study the Levenberg-Marquardt error propagation algorithm was used due to higher convergence than other training functions, and the condition for stopping neural network training is an error coefficient of 0.001.

## 3. Block diagram of proposed method

**Figure 8** shows the block diagram of the steps of the proposed method to diagnose Alzheimer's disease.

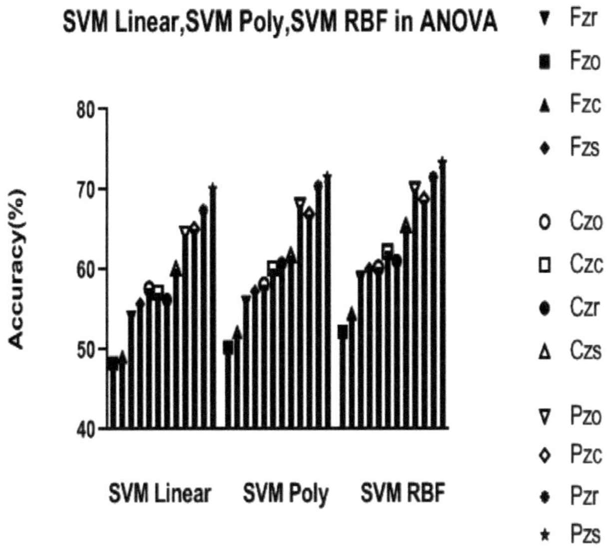


**Figure 8.**
*Results of SVM signal separation accuracy with different core functions with optimal characteristics obtained. from ANOVA.*

## 4. Result

Selected optimal features by ANOVA method in four modes of the closed eye, open eye, Reminder and stimulation are shown in **Table 2** to compare the number and types of optimum features between Fz, Cz, Pz channels. The results of the accuracy of the separation of brain signals using SVM with different core functions with optimal characteristics obtained from ANOVA are shown in **Figure 8**. **Figure 9** evaluates closed-eye, open-eye, reminder, and stimulation modes for the desired channels. According to the four modes of closing the eyes, opening the eyes, reminding and stimulating the brain signal, in order to better identify and introduce the features more accurately, each section is considered with a certain index. The closed part is considered with index c, the open eye part is considered with index o, the reminder part is with index r, and the stimulation part is considered

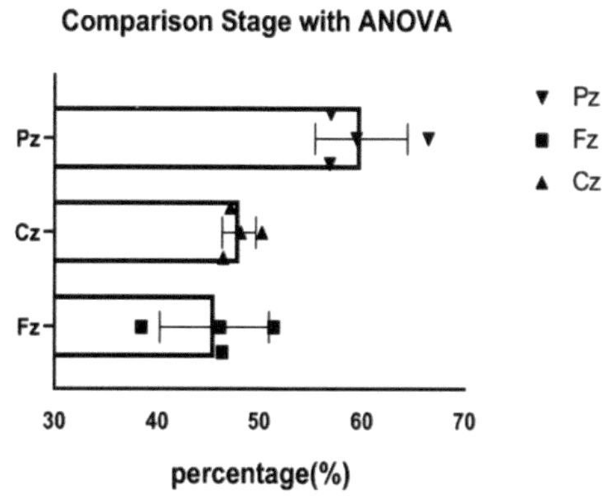


**Figure 9.**
*Evaluation of closed-eye, open-eye, reminder and excitation modes for Fz, Cz, Pz channels in ANOVA mode.*

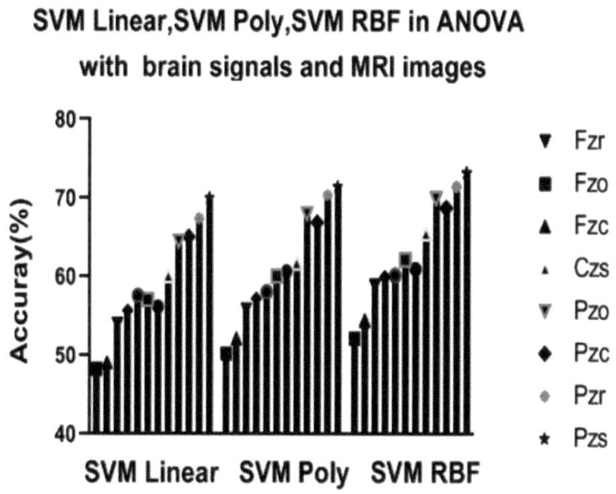


**Figure 10.**
*Compares the results of the resolution of brain signals by the SVM classifier with different core functions with ANOVA-optimized features with MRI features and without MRI features.*

with index s. In the excitation section, the target and non-target sound sections are defined with st and ss indices, respectively. In **Figure 10**, the results of separation of brain signals by SVM classifier with different main functions are compared with optimized ANOVA features with MRI features and without MRI features.

Due to the comparison of the results in **Figure 11**, by the support vector machine with different cores by the optimal brain signal characteristics by ANOVA with the addition of MRI image features, the accuracy of the results is reduced compared to the case where only the brain signal features are used.

Finally, by using neural network dynamics because of the nonlinear properties studied and due to the nonlinear dynamics of the EEG signal, Elman neural network is used. The results in **Figure 12** are compared by Elman with the optimal features of the brain signal obtained from ANOVA and with the addition of the features of MRI images, and the accuracy of the results is increased compared to the case where only the features of the brain signal are used.

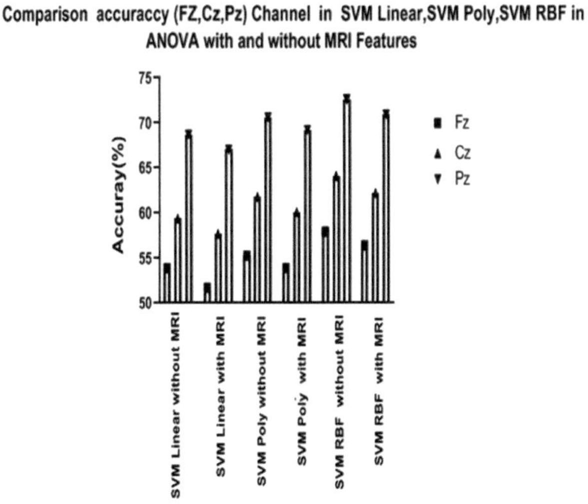


**Figure 11.**
*Support vector machine with different cores by the optimal brain signal characteristics by ANOVA with the addition of MRI image.*

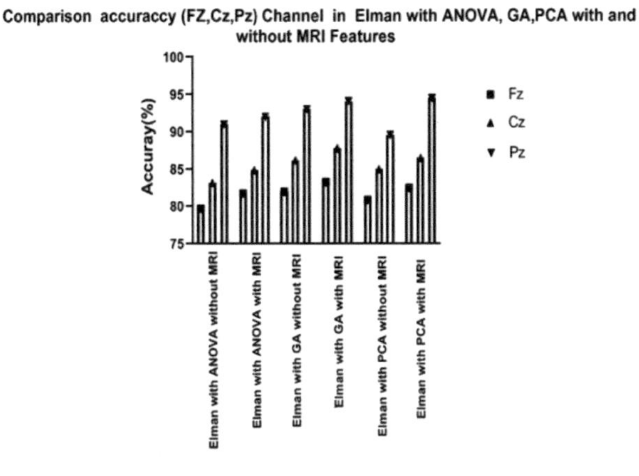


**Figure 12.**
*Compared by Elman with the optimal features of the brain signal obtained from ANOVA and with the addition of the features of MRI images.*

## 5. Discussion

Medical image analysis has become very important in the diagnosis of mild Alzheimer's disease in recent years [3]. But the important point is that from the way of analyzing medical images, we can determine the most effective channel for recording brain signals. 3D segmentation of MRI images further helps researchers diagnose Alzheimer's disease and obtain important information [7]. And in 3D images, the most appropriate direction in the image is effective in determining the appropriate features. Determining the degree of atrophy of MRI images is an effective method for early detection of Alzheimer's disease. Also, assessing the degree of asymmetry in both the right and left hemispheres and analyzing volumetric mismatch can differentiate between mild and severe Alzheimer's disease [7]. The degree of asymmetry in the left and right hemispheres should be determined by the degree of atrophy and the ratio of the volume of gray matter to the volume of white matter. Using statistical features of signal and obtaining temporal information and using spatial features of MRI images is an effective method for the more accurate evaluation of Alzheimer's disease [8]. The statistical properties of the signal are temporal in nature and the statistical properties of the image are spatial in nature. Cortical atrophy means the gradual destruction of the nerve cells that make up the upper regions of the brain, specifically the structures found in the cerebral cortex, mostly due to a reduction or loss of oxygen and nutrients in these areas. There are several methods for examining the Medial temporal lobe, the accuracy of which is not clear [9]. However, this condition is more suitable for mild patients. Longitudinal T1-weighted MRI studies are another effective way to distinguish mild Alzheimer's patients from healthy ones [10]. But this feature has better results in severe patients to differentiate with mild patients. Also, extracting the appropriate characteristics and deciding on the classification in this field are among the issues to be considered.

## 6. Conclusion

Investigation and analysis of nonlinear dynamics of brain signal show nonlinear and dynamic behavior in different stages of Alzheimer's disease. Nonlinear dynamics analysis of this signal shows a decrease in the complexity of the brain signal pattern and a decrease in connections due to a decrease in the nonlinear cell dynamics between cortical regions. The next two features are correlation and Lyapanov's appearance, which indicates the feature space, and the convergence or divergence of this space is slightly reduced in this disease. The courses studied are closed-eyed, open-eyed, reminder, and stimulation, and among these four periods, the stimulation period was the best period for recording brain signals, because to diagnose Alzheimer's disease, it is more effective to evaluate the speed of stimulus-response. The mean rate of asymmetry and the mean rate of temporal lobe atrophy increase with the progression of Alzheimer's disease because the amount of damage to the temporal lobe in MRI images of Alzheimer's disease has increased. The accuracy of the results in Elman neural network with the combination of brain signal features and medical images is 94.4% and in the case without combining the signal and image features, the accuracy of the results is 92.2%. The use of nonlinear classifiers is more appropriate than other classification methods due to the nonlinear dynamics of the brain signal. The accuracy of the results in the support vector machine with RBF core with the combination of brain signal features and medical images is 75.5% and in the case without combining the signal and image features, the accuracy of the results is 76.8%. Due to its nonlinear and normal distribution nature, this nucleus has been able to produce better results. Among the processing methods proposed to classify the three classes of healthy, mild, and severely ill, the method of combining brain signal characteristics and medical images has increased the accuracy of Elman classifier results and decreased the accuracy of SVM results. Because spatial features do not have the same nature as temporal features, and if the classifier divides the groups based on linear and non-return methods by extracting inappropriate features, the correct results will not be created. The main innovation in this research is the extraction of the most appropriate features and the appropriate combination of spatial features of medical images and temporal features of brain signals to diagnose Alzheimer's disease.

## Author details

Elias Mazrooei Rad
Department of Biomedical Engineering, Khavaran Institute of Higher Education, Mashhad, Iran

*Address all correspondence to: elias_mazrooei@yahoo.com

## References

[1] Roselli F, Tartaglione B, Federico F, Lepore V, Defazio G, Livrea P. Rate of MMSE score change in Alzheimer's disease: Influence of education and vascular risk factors. Clinical Neurology and Neurosurgery. 2009;**111**(4):327-330

[2] Prince MJ, Wimo A, Guerchet MM, Ali GC, Wu YT, Prina M. World Alzheimer Report 2015-The Global Impact of Dementia: An analysis of prevalence, incidence, cost and trends, 2015

[3] Biju KS, Alfa SS, Lal K, Antony A, Akhil MK. Alzheimer's detection based on segmentation of MRI image. Procedia Computer Science. 2017;**115**:474-481

[4] Zhao X, Ang CK, Acharya UR, Cheong KH. Application of artificial intelligence techniques for the detection of Alzheimer's disease using structural MRI images. Biocybernetics and Biomedical Engineering. 2021;**41**(2):456-473

[5] El-Sappagh S, Alonso JM, Islam SR, Sultan AM, Kwak KS. A multilayer multimodal detection and prediction model based on explainable artificial intelligence for Alzheimer's disease. Scientific Reports. 2021;**11**(1):1-26

[6] Hett K, Ta VT, Catheline G, Tourdias T, Manjón JV, Coupé P. Multimodal hippocampal subfield grading for Alzheimer's disease classification. Scientific Reports. 2019;**9**(1):1-6

[7] Clerx L, van Rossum IA, Burns L, Knol DL, Scheltens P, Verhey F, et al. Measurements of medial temporal lobe atrophy for prediction of Alzheimer's disease in subjects with mild cognitive impairment. Neurobiology of Aging. 2013;**34**(8):2003-2013

[8] Huang A, Abugharbieh R, Tam R. A hybrid geometric–statistical deformable model for automated 3-D segmentation in brain MRI. IEEE Transactions on Biomedical Engineering. 2009;**56**(7):1838-1848

[9] Visser PJ, Verhey FR, Hofman PA, Scheltens P, Jolles J. Medial temporal lobe atrophy predicts Alzheimer's disease in patients with minor cognitive impairment. Journal of Neurology, Neurosurgery & Psychiatry. 2002;**72**(4):491-497

[10] Sun Z, van de Giessen M, Lelieveldt BP, Staring M. Detection of conversion from mild cognitive impairment to Alzheimer's disease using longitudinal brain MRI. Frontiers in Neuroinformatics. 2017;**11**:16

[11] Jiang X, Bian GB, Tian Z. Removal of artifacts from EEG signals: A review. Sensors. 2019;**19**(5):987

[12] Micanovic C, Pal S. The diagnostic utility of EEG in early-onset dementia: A systematic review of the literature with narrative analysis. Journal of Neural Transmission. 2014;**121**(1):59-69

[13] Szirmai I, Kamondi A. EEG investigations in cognitive impairments. Ideggyogyaszati szemle. 2011;**64**(1-2):14-23

[14] Jackson CE, Snyder PJ. Electroencephalography and event-related potentials as biomarkers of mild cognitive impairment and mild Alzheimer's disease. Alzheimer's & Dementia. 2008;**4**(1):S137-S143

[15] Lee MS, Lee SH, Moon EO, Moon YJ, Kim S, Kim SH, et al. Neuropsychological correlates of the P300 in patients with Alzheimer's disease. Progress

in Neuro-Psychopharmacology and Biological Psychiatry. 2013;**40**:62-69

[16] Burton EJ, Barber R, Mukaetova-Ladinska EB, Robson J, Perry RH, Jaros E, et al. Medial temporal lobe atrophy on MRI differentiates Alzheimer's disease from dementia with Lewy bodies and vascular cognitive impairment: A prospective study with pathological verification of diagnosis. Brain. 2009;**132**(1):195-203

[17] Hajmanouchehri R. CT scan and MRI findings in patients with dementia. Scientific Journal of Forensic Medicine. 2017;**23**(3):150-159

[18] Wood PL, Barnette BL, Kaye JA, Quinn JF, Woltjer RL. Non-targeted lipidomics of CSF and frontal cortex grey and white matter in control, mild cognitive impairment, and Alzheimer's disease subjects. Acta Neuropsychiatrica. 2015;**27**(5):270-278

[19] Brill FZ, Brown DE, Martin WN. Fast generic selection of features for neural network classifiers. IEEE Transactions on Neural Networks. 1992;**3**(2):324-328

[20] Chowdhury RH, Reaz MB, Ali MA, Bakar AA, Chellappan K, Chang TG. Surface electromyography signal processing and classification techniques. Sensors. 2013;**13**(9):12431-12466

[21] Rabeh AB, Benzarti F, Amiri H. Diagnosis of alzheimer diseases in early step using SVM (support vector machine). In: 2016 13th International Conference on Computer GraphicsI Imaging and Visualization (CGiV). IEEE; 2016. pp. 364-367

[22] Papadaniil CD, Kosmidou VE, Tsolaki A, Tsolaki M, Kompatsiaris IY, Hadjileontiadis LJ. Cognitive MMN and P300 in mild cognitive impairment and Alzheimer's disease: A high density EEG-3D vector field tomography approach. Brain Research. 2016;**1648**:425-433

Chapter 4

# Saliency Detection from Subitizing Processing

*Carola Figueroa-Flores*


## Abstract

Most of the saliency methods are evaluated for their ability to generate saliency maps, and not for their functionality in a complete vision pipeline, for instance, image classification or salient object subitizing. In this work, we introduce saliency subitizing as the weak supervision. This task is inspired by the ability of people to quickly and accurately identify the number of items within the subitizing range (e.g., 1 to 4 different types of things). This means that the subitizing information will tell us the number of featured objects in a given image. To this end, we propose a saliency subitizing process (SSP) as a first approximation to learn saliency detection, without the need for any unsupervised methods or some random seeds. We conduct extensive experiments on two benchmark datasets (Toronto and SID4VAM). The experimental results show that our method outperforms other weakly supervised methods and even performs comparable to some fully supervised methods as a first approximation.


**Keywords:** saliency prediction, subitizing, object recognition, deep learning and convolutional neural network

## 1. Introduction

For humans, object recognition is a nearly instantaneous, precise, and extremely adaptable process. Furthermore, it has the innate ability to learn new classes of objects from a few examples [1, 2]. The human brain reduces the complexity of incoming data by filtering out some of the information and processes only those things that grab our attention. This, combined with our biological predisposition to respond to certain shapes or colors, allows us to recognize at a glance the most important or outstanding regions of an image. This mechanism can be observed by analyzing which parts of the images humans pay more attention to; for example, where they fix their eyes when they are shown an image [3, 4]. The most accurate way to record this behavior is by tracking eye movements, while the subject in question is presented with a set of images to evaluate. Computational estimation of saliency (or salient or salient regions) aims to identify to what extent regions or objects stand out from their surroundings or background to human observers. Saliency maps can be used in a wide range of applications, including object detection, image and video understanding, and eye tracking. On the other hand, it is known that the human visual system can effortlessly identify the number of objects in the range 1 to 4 by having just one glance [5]. Since then, this phenomenon, coined later by [6] as subitizing, has been studied and tested in various experimental settings [7].

Therefore, inspired by subitizing and the results obtained in [8, 9], the main objective of this project is to incorporate the subitizing of salient objects (SOS), in order to improve our previous results. This means that the subitizing information will tell us the number of outgoing objects in a given image and thus subsequently provide us with the location or appearance information of the outgoing objects explicitly, and everything will be done within a weakly supervised configuration. It should be noted that when the network is trained with the subitizing supervisions, the network will learn to focus on the regions related to the outgoing objects. Therefore, it will design a saliency subitizing process (SSP) architecture that is responsible for extracting attention regions as saliency map. A second module that is in charge of improving the quality of the saliency masks can be defined as the saliency map update process (SUP), which will basically be in charge of refining the activation regions in an end-to-end way. It will then merge the source images and saliency maps to get the masked images as new inputs for the next refinement. Finally, in this work we propose to design and build a convolutional neural network (CNN), which will basically consist of a process that will be in charge of SSP and a function that will help us in the task of SUP. The first SSP will serve as a support to obtain and calculate the number of outstanding objects and thus extract the saliency maps with their respective locations. Instead, SUP will help us update the saliency masks produced by the first module. The general model of our proposal is shown in **Figure 1**.

However, as this work is a first attempt at the final result, it will only consider the development, experimentation, and explanation associated with step 1.

It briefly summarizes below its main contributions:

- It proposes an approach that generates saliency maps from subitizing of saliency process (SSP),

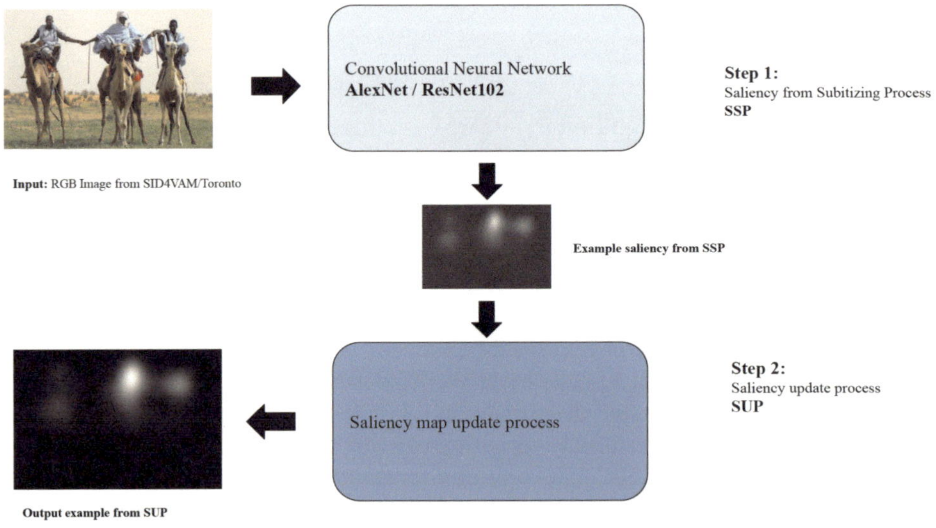


**Figure 1.**
*Overview of our proposed method with the saliency subitizing process (SSP) and the saliency updating process (SUP).*

- Our saliency does not require any saliency maps for training (like previous works [10, 11]) but instead is trained indirectly in an end-to-end fashion by training the network for image classification with subitizing saliency process (SSP).

- The saliency maps obtained without using any saliency groundtruth data shows competitive results.

The chapter is organized as follows. Section 2 is devoted to review the related work in saliency detection. Section 3 presents our approach. Experimental results are reported in Section 4. Finally, Section 5 contains our conclusions.

## 2. Related work

Saliency is generally known as local contrast [12], which generally originates from contrasts between objects and their surroundings, such as differences in color, texture, shape. This mechanism measures intrinsically salient stimuli to the vision system that primarily attracts the attention of humans, in the initial stage of visual exposure to an input image [13]. To quickly extract the most relevant information from a scene, the human visual system pays more attention to highlighted regions, as seen in **Figure 1**. Research on computational saliency focuses on the design of algorithms that, like human vision, predict which regions of a scene stand out [14, 15].

Initial efforts to model saliency involved multi-scale representations of color, orientation, and intensity contrast. These were often biologically inspired, such as the well-known works [12, 16]. From that model, a large number of models were based on the manual elaboration of these features to obtain an accurate saliency map [17, 18], either maximizing [19] or learning statistics from natural images [13, 20]. Relevancy research was further driven by the availability of large datasets that enabled the use of machine learning algorithms [21], primarily pre-trained on existing human fixation data.

The question of whether saliency is important for object recognition and tracking has been raised in [22]. More recent methods [23] take advantage of end-to-end convolutional architectures by fine graining on fixation prediction [4, 24, 25]. But the main goal of these works was to estimate a saliency map, not how saliency might contribute to object recognition.

Several works have shown that having the saliency map of an image can be useful for object recognition, for example, [8, 10, 11]. Since the saliency map can help focus attention on the relevant parts of the image to improve recognition, additionally, it can help guide training by focusing backpropagation on relevant image regions. Previous work has shown that saliency modulated image classification (SMIC) is especially efficient for training on data sets with few labeled data [10]. The main drawback of these methods is that they require a trained saliency method. Also, Refs. [8, 9] show that this restriction can be removed and that it can hallucinate the saliency image from the RGB image. By training the network for image classification on the ImageNet dataset [26], it can obtain the saliency branch without using human reference images.

Recently, the progress in the detection of salient objects has grown substantially, mainly benefiting from the development of deep neural networks (CNN). In [27], a CNN based on the use of superpixels for saliency detection was proposed. Instead, Li et al. [28] used multi-scale features extracted from a deep CNN. Zhao et al. [29] proposed a multi-context deep learning framework to detect salient objects with two different CNNs, which were useful for learning local and global information. Yuan et al. [30] proposed a saliency detection framework, which extracted the correlations between object contours and RGB features of the image. On the other hand, Wang and Shen [31] defined a pyramid-shaped structure to expand the receptive field in visual attention. Hou and Zhang [32] introduced short connections for edge or contour detection. Zhu [33], on the other hand, proposed a visual attention architecture called DenseASPP, to extract information. Chen [34] proposed a spatial attenuation context network, which recursively translated and aggregated the context features in different layers. Tu [35] introduced an edge-guided block to embed boundary information in saliency maps. Zhou [36] proposed a multi-type self-attention network to learn more semantic details from degraded images. However, these methods rely heavily on pixel-based monitoring. Overcoming the scarcity of pixel-based data, it focusses on the saliency detection task.

### 2.1 Weakly supervised saliency detection

There are many works using weak supervisions for the saliency detection task. For example, Li [37] used the image-level labels to train the classification network and applied coarse activation maps as saliency maps. Wang [38] proposed a weakly supervised two-stage method by designing an inference network to predict foreground regions and global smooth pooling (GSP) to aggregate responses from those predicted objects. On the other hand, Zeng [39] designed a unified network, which is capable of weak monitoring of multiple sources, including image labels, captions, and pseudo-labels. Furthermore, they designed a loss of attention transfer to transmit signals between subnetworks with different supervisions.

Different from the previous methods, it proposes to use subitizing information as weak supervision in the saliency detection task, where it will first study the problem of subitizing of the outgoing object and the relationships between subitizing and saliency detection.

## 3. Proposed method

This work proposes to design and implement a convolutional neural network, which will consist mainly of saliency subitizing process (SSP). The SSP will help us to count the highlighted objects and at the same time extract the saliency from the maps that will contain the locations (positions) of the objects.

### 3.1 Subitizing of saliency process (SSP)

It should be noted that the information provided by the subitizing process will indicate the number of outgoing objects in a given image [40]. Therefore, it will not explicitly provide the location or information related to the appearance of the output objects. However, when the network is being trained with subitizing (simulating supervised learning), the network will learn to focus on the regions related to the most salient (or salient) objects. Training images are divided into five categories based on the number of salient objects: 0, 1, 2, 3, and 4 or +. For the same reason, it will design the SSP to extract these regions as if it were a saliency mask. During this process, a classification network will be used for the object subitizing task, in this context ResNet-152 or ResNet50 [41] and AlexNet [42] as "backbone network," which are pre-trained from the ImageNet dataset [43].

Also, it uses cross-entropy as the classification loss (see Eq. (1)). In order to obtain denser saliency maps, the stride of the last two down-sampling layers is set as 1 in our backbone network, which produces feature maps with 1/8 of the original resolution before the classification layer. In order to enhance the representation power of the proposed network, it also applies two attention modules: channel attention module and spatial attention module, which tell the network "where" and "what" to focus, respectively. Both of them are placed in a sequential way between the ResNet blocks and AlexNet convolutional layers.

$$\mathcal{I} = \sum_{I \in \mathcal{D}} logp_{c(I)}(y|I) \tag{1}$$

In addition, it applies the technique of the gradient-weighted class activation mapping (Grad-CAM) [44] to extract salient regions as the initial saliency maps, which contains the gradient information flowing into the last convolutional layers during the backward phase. The gradient information represents the importance of each neuron during inference of the network. It assumes that the features produced from the last convolutional layer has a channel size of $K$. For a given image, let $f_k$ be the activation of unit $K$, where $k \in [1, K]$. For each class $c$, the gradients of the score $y^c$ with respect to activation map $f_k$ are averaged to obtain the neuron significant weight $a_k^c$ of class $c$:

$$a_k^c = \frac{1}{N} \sum_i^m \sum_j^h \frac{\partial y^c}{\partial f_{i,j}^k} \tag{2}$$

where $i$ and $j$ represent the coordinates in the features map $N = m \times h$. With the neuron importance weight $a_k^c$, we can compute the activation map $M^c$:

$$M^c = ReLU\left(\sum_k a_k^c f^k\right) \tag{3}$$

And, finally it adds an activation map with ReLU (rectified linear unit) function layer; this function filters negative gradient values, since only the positive ones contribute to the class decision, while the negative values contribute to other categories. The size of the saliency map is the same as the size of the last convolutional feature maps (1/8 of the original resolution). This process is shown in **Figure 2**.

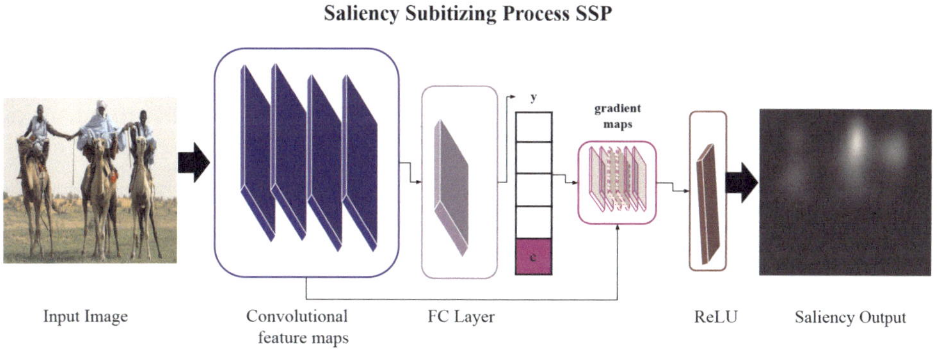


**Figure 2.**
*The pipeline of the saliency subitizing process (SSP).*

## 4. Experiments

This section discusses the advantage of SSP that help us to learn counting of salient objects and extract coarse saliency maps with the precise locations of the target objects.

### 4.1 Experimental setup

**Datasets.** The saliency maps have been computed for images from a distinct eye-tracking dataset, corresponding to 120 real scenes (Toronto) [19] and 230 synthetic images with specific feature contrast (SID4VAM) (see **Table 1**) [45]. These images datasets have been computed with our approach, a supervised artificial model that specifically computes high-level features (DeepGazeII, ML-Net (multi-level net), SAM (saliency attentive model), salGAN), and models biological inspiration (IKN (Itti, Koch, and Niebur) [16], AIM [19] (saliency based on information maximization), SDLF (saliency detection by using local features) [20], and GBVS (graph-based visual saliency) [13]).

**Networks architectures.** It evaluates approach using two network architectures: AlexNet [42] and ResNet-152 [41]. It is modified to meet our requirement. In both cases, the weights were pretrained on ImageNet and then fine-tuned on each of the datasets mentioned above. The networks were trained for 70 epochs with a learning rate of 0.0001 and a weight decay of 0.005. The top classification layer was initialized from scratch using Xavier method [46]. The SSP consists of four convolutional layers for AlexNet and four residual blocks for ResNet-152.

**Comparison.** This work compares its proposal with other models (see **Tables 2** and **3**—rows 8) from fixation data. For instance, DeepGazeII summed the center baseline, whereas in ML-Net and SAM, the learned priors are used for modulating the result of the network.

### 4.2 Results

#### *4.2.1 First experiment: Multiple networks*

In order to evaluate how accurately the saliency map is able to match the location of human fixations, it used a set of metrics previously defined by [17].

In **Table 4** we show results of area under ROC (AUC), correlation coefficient (CC), normalized scanpath saliency (NSS), Kullback-Leibler divergence (KL), and similarity (SIM) for every network for all datasets.

The area under ROC (AUC) is considered as true positives, the saliency map values coincide with a fixation and false positives, and the saliency map values that have no fixation then compute the area under the curve. Similarly, the NSS computes the

| Data Set | Type | # Images | # PP | pxva | Resolution |
|---|---|---|---|---|---|
| Toronto | Indoors and outdoors | 120 | 20 | 32 | 681x511 |
| SID4VAM | Synthetic pop-out | 230 | 34 | 40 | 1280x1024 |

*pxva: pixels per 1 degree of visual angle, PP: participants.*

**Table 1.**
*Characteristics of eye-tracking datasets.*

| Method | AUC | KL ↓ | SIM | sAUC | InfoGain |
|---|---|---|---|---|---|
| IKN [16] | *0.782* | *1.249* | 0.366 | 0.650 | −0.024 |
| AIM [19] | 0.716 | 1.612 | 0.314 | 0.663 | −0.580 |
| SDLF [20] | 0.703 | 1.518 | 0.304 | 0.664 | −0.398 |
| GBVS [13] | *0.803* | *1.168* | 0.397 | 0.632 | *0.077* |
| DeepGazeII [24] | ***0.838*** | 1.367 | 0.325 | 0.763 | −0.200 |
| SAM-ResNet [4] | 0.725 | 2.420 | ***0.516*** | *0.666* | −1.555 |
| SalGAN [47] | *0.818* | 1.272 | *0.435* | *0.715* | ***0.392*** |
| Our Approach (SSP) | 0.740 | 1.409 | 0.399 | 0.597 | −0.399 |
| *GroundTruth (Humans)* | *0.954* | *0.000* | *1.000* | *0.902* | *2.425* |

**Table 2.**
*Comparison of our saliency output with standard benchmark methods over real image* ***Toronto*** *dataset for saliency prediction. (Top) Baseline low-level saliency models. (Bottom) State-of-the-art deep saliency models. Best score for each metric is defined as* ***bold*** *and TOP-3 scores are italicized.*

| Method | AUC | KL ↓ | SIM | sAUC | InfoGain |
|---|---|---|---|---|---|
| IKN [16] | *0.678* | 1.748 | 0.380 | 0.608 | −0.233 |
| AIM [19] | 0.566 | 14.472 | 0.224 | 0.557 | −18.181 |
| SDLF [20] | 0.607 | 3.954 | 0.322 | 0.596 | −3.244 |
| GBVS [13] | *0.718* | ***1.363*** | *0.413* | *0.628* | ***0.331*** |
| DeepGazeII [24] | 0.610 | *1.434* | 0.335 | 0.571 | −0.964 |
| SAM-ResNet [4] | 0.673 | 2.610 | *0.388* | 0.600 | −1.475 |
| SalGAN [47] | 0.662 | 2.506 | 0.373 | 0.593 | −1.350 |
| Our Approach (SSP) | ***0.741*** | 1.658 | ***0.445*** | ***0.633*** | *−0.122* |
| *GroundTruth (Humans)* | *0.882* | *0.000* | *1.000* | *0.860* | *2.802* |

**Table 3.**
*Comparison of our saliency output with standard benchmark methods over synthetic image SID4VAM dataset for saliency prediction. (Top) Baseline low-level saliency models. (Bottom) State-of-the-art deep saliency models. Best score for each metric is defined as bold and TOP-3 scores are italicized.*

| Dataset | Model | AUC-Judd | AUC-Borji | CC | NSS | KL↓ | SIM |
|---|---|---|---|---|---|---|---|
| | AlexNet | 0.7655 | 0.7298 | 0.4603 | 1.3888 | *1.5155* | 0.3955 |
| Toronto | ResNet152 | *0.7911* | *0.7443* | *0.5440* | *1.6391* | 1.6891 | *0.4410* |
| | AlexNet | 0.6910 | 0.7366 | 0.3889 | 1.4106 | *1.7152* | *0.4385* |
| SID4VAM | ResNet152 | *0.7015* | *0.7723* | *0.3910* | *1.1155* | 1.9890 | 0.3996 |

**Table 4.**
*Benchmark of our method with different networks (top 1 networks are italicized).*

average normalized saliency map that coincides with fixations. Other metrics such as CC, KL, and SIM compute the score upon the region distribution statistics of all pixels (KL calculating the divergence and CC/SIM the histogram intersection or similarity of the distribution).

After computing the saliency maps for all datasets (see in **Table 4**) with AlexNet and ResNet152, we observed that metric scores vary considerately depending on dataset or network. AlexNet is shown to provide best results for pop-out patterns (SID4VAM), whereas ResNet152 shows overall higher scores with real images of Toronto dataset.

*4.2.2 Second experiment: Qualitative results*

These saliency prediction results show that our model has robust metric scores on both real and synthetic images for saliency prediction. Again, we would like to stress that our model is not trained on fixation prediction datasets (**Figure 3**). Its model with subitizing supervision performs best on detecting pop-out effects (from visual attention theories [16]) while performing similarly for real image datasets (**Figure 4**). Some deep saliency models use several mechanisms to leverage (or/and train) performance for improving saliency metric scores, such as smoothing/thresholding (see **Figure 4**, row 4). It is also considered that some of these models are already fine-tuned for synthetic images (e.g., SAM-ResNet [4]). *Our approach* (which has not been trained in these type of data sets) has shown to be robust on these two distinct scenarios/domains.

### 4.3 Evaluation benchmark of saliency estimation

Here, we compare the saliency estimation that is obtained after only performing Step 1 in **Figure 1** with existing saliency models (see **Table 5**). This saliency estimation is trained without access to any groundtruth saliency data.

Saliency prediction metrics assign a score depending on how well the predicted saliency map is able to match with locations of human fixations (see definitions in Borji et al. [17]). It selected the area under ROC (AUC), Kullback-Leibler divergence (KL), similarity (SIM), shuffled AUC (sAUC) and information gain (IG) metrics considering its consistency of predictions of human fixation maps. It compares scores

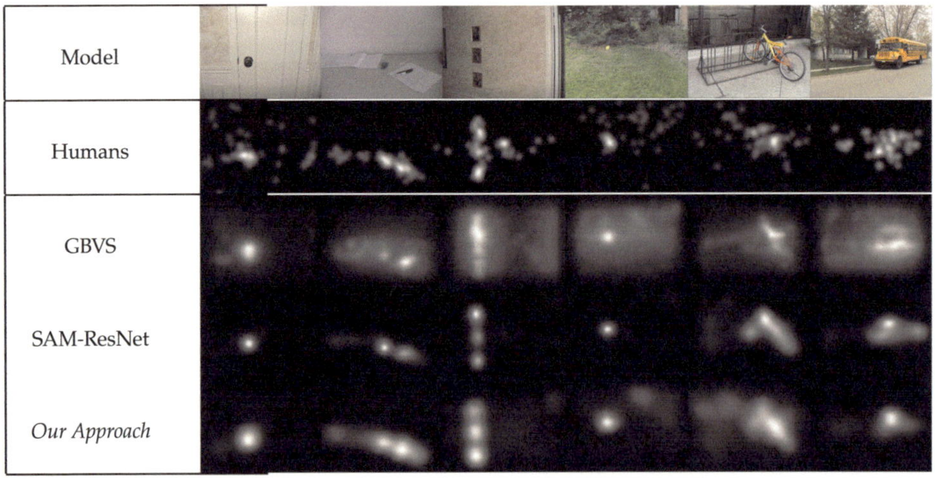


**Figure 3.**
*Qualitative results for real images (Toronto dataset). Each image is represented in a different column and each model saliency map in each row. The ground truth density map of human fixations is represented in the second row.*

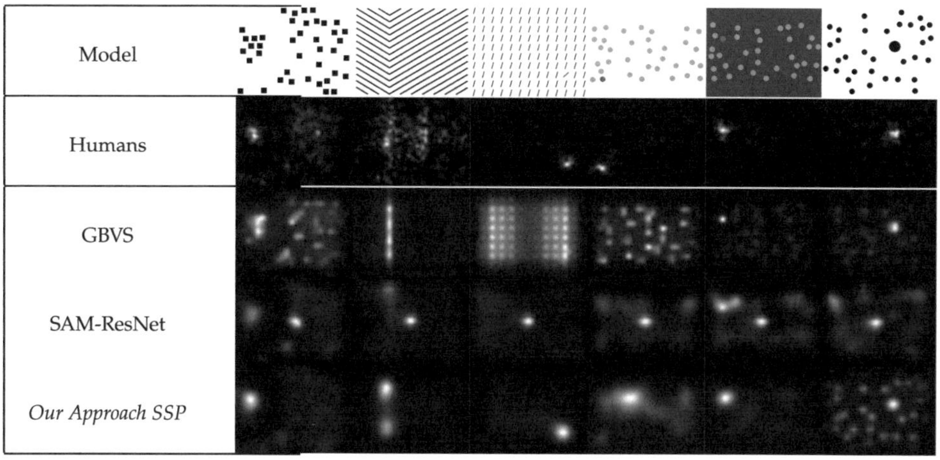


**Figure 4.**
*Qualitative results for synthetic images (SID4VAM dataset). Each image is represented in a different column and each model saliency map in each row. The ground truth density map of human fixations is represented in the second row.*

with classical saliency models, both with handcrafted low-level features (i.e., IKN [16], AIM [19], SDLF [20], and GBVS [13]) and with state-of-the-art deep saliency models (i.e., DeepGazeII [24], SAM-ResNet [4], and SalGAN [47]) mainly pretrained on human fixations. The results are surprising; our method, which has not been trained on any saliency data, obtains competitive results. For the case of *Toronto* (**Table 2**), the best model is GBVS, followed by our model, which scores in the top 3 of KL and SAM-ResNet and scores slightly higher in InfoGain metric. For the case of *SID4VAM* (**Table 3**), our approach gets the best scores for most metrics compared with other deep saliency models, being mainly among the top 2 acquiring similar scores to GBVS in most metrics (outperforming it in AUC measures).

These saliency prediction results show that our model has robust metric scores on both real and synthetic images for saliency prediction. Again, we would like to stress that our model is not trained on fixation prediction datasets and our model with subitizing supervision (SUP) performs best on detecting pop-out effects (from visual attention theories [16]), while performing similarly for real image datasets (**Figure 4**). Some deep saliency models use several mechanisms to leverage (or/and train) performance for improving saliency metric scores, such as smoothing/thresholding (see **Figure 4**, rows 5). It also considers that some of these models are already fine-tuned for synthetic images (e.g., SAM-ResNet [4]). *Our approach* (which has not been trained in these types of datasets) has shown to be robust on these two distinct scenarios/domains.

## 5. Conclusions

In this chapter, we proposed a method for the saliency estimation with weak subitizing supervision. We designed a model with the saliency subitizing process (SSP), which generates the initial saliency map using subitizing information. Without any seeds from unsupervised methods, this method outperforms other weakly supervised methods and even performs comparable to some fully supervised methods.

| # | Name | Year | Features/Architecture | Mechanism |
|---|---|---|---|---|
| 1 | IKN | 1998 | DoG (color+intensity) | — |
| 2 | AIM | 2005 | ICA (infomax) | max-like |
| 3 | GBVS | 2006 | Markov chains | graph prob. |
| 4 | SDLF | 2006 | Steerable pyramid | local+global prob. |
| 5 | ML-Net | 2016 | VGG-16 | Backprop.(finetuning) |
| 6 | DeepGazeII | 2016 | VGG-19 | Backprop.(finetuning) |
| 7 | SAM | 2018 | VGG-16/ResNet-50 + LSTM | Backprop.(finetuning) |
| 8 | SalGAN | 2017 | VGG-16 Autoencoder | Finetuning+GAN Loss |
| # | Name | Learning | Training Data (#img) | Bias/Priors |
| 1 | IKN | — | — | — |
| 2 | AIM | Unsupervised | Corel (3600) | — |
| 3 | GBVS | Unsupervised | Einhauser (108) | graph norm. |
| 4 | SDLF | Unsupervised | Oliva (8100) | scene priors |
| 5 | ML-Net | SALICON (10 k), MIT (1003) | learned priors | — |
| 6 | DeepGazeII | Supervised | SALICON (10 k), MIT (1003) | center bias |
| 7 | SAM | Supervised | SALICON (10 k) & others | Gaussian priors |
| 8 | SalGAN | Supervised | SALICON (10 k), MIT (1003) | — |

*DoG: difference of Gaussians, ICA: independent component analysis, C-S: center-surround, max-like: max-likelihood probability, BCE: binary cross-entropy, GAN: generative adversarial network.*

**Table 5.**
*Description of saliency models.*

Finally, as this work is a first approximation, future work would be to verify how its saliency map would improve if the SUP update module were added.

## Acknowledgements

We thank the support from FOVI21001 "Fomento a la Vinculación Internacional para Instituciones de Investigación Regionales (ANID, Chile)," Agencia Nacional de Investigación y Desarrollo and ALBA Research Group (Algorithms and Database) 2130591 GI/VC, "Ayudantes para el Fortalecimiento de Investigación FACE 2022," and "Proyecto de Reinserción" DIUBB 2230508 IF/RS of the University of Bío.

## Additional information

This chapter is a continuation of my PhD thesis, previous works related to the subject of saliency, and the use of the subitizing technique, because it had already been tested by other works, and in this way, the saliency estimation of my previous works was improved.

## Author details

Carola Figueroa-Flores
Department of Computer Science and Information Technology, Universidad del Bío Bío, Chile

*Address all correspondence to: cfigueroa@ubiobio.cl

## References

[1] Sun X, Yao H, Ji R, Liu XM. Toward statistical modeling of saccadic eye-movement and visual saliency. IEEE Transactions on Image Processing. 2014; **23**(11):4649-4662

[2] Vincent BT, Tatler BW. Systematic tendencies in scene viewing. Journal of Eye Movement Research. 2008:1-18. eyemovement.org. DOI: 10.16910/jemr.2.2.5

[3] Achanta R, Hemami S, Estrada F, Susstrunk S. Frequency-tuned salient region detection. In: 2009 IEEE Conference on Computer Vision and Pattern Recognition. 2009. pp. 1597–1604

[4] Cornia M, Baraldi L, Serra G, Cucchiara R. Predicting human eye fixations via an LSTM-based saliency attentive model. IEEE Transactions on Image Processing. 2018b;**27**(10):5142-5154

[5] Stanley J. The power of numerical discrimination. Nature. 1871;**3**:367-367. DOI: 10.1038/003367b0

[6] Kaufman EL, Lord Miles W, Whelan RT, Volkmann J. The discrimination of visual number. The American Journal of Psychology. 1949;**62**:498-525

[7] Whalen J, Gallistel CR, Gelman R. Nonverbal counting in humans: The psychophysics of number representation. Psychological Science. 1999;**10**:130-137. DOI: 10.1111/1467-9280.00120

[8] Flores CF, Raducanu BC, Berga D, van de Weijer J. Hallucinating saliency maps for fine-grained image classification for limited data domains. VISIGRAPP (4: VISAPP). 2021. pp. 163-171

[9] Figueroa-Flores C, Berga D, van de Weijer J, Raducanu B. Saliency for free: Saliency prediction as a side-effect of object recognition. Pattern Recognition Letters. 2021:1-7. DOI: 10.1016/j.patrec.2021.05.015

[10] Figueroa-Flores C, Gonzalez-Garcia A, van de Weijer J, Raducanu B. Saliency for fine-grained object recognition in domains with scarce training data. Pattern Recognition. 2019;**94**:62-73

[11] Murabito F, Spampinato C, Palazzo S, Giordano D, Pogorelov K, Riegler M. Top-down saliency detection driven by visual classification. Computer Vision and Image. 2018:67-76. Understanding

[12] Itti L, Koch C. Computational modeling of visual attention. Nature Reviews. Neuroscience. 2001;**2**:194-203. DOI: 10.1038/35058500

[13] Harel J, Koch C, Perona P. Graph-based visual saliency. In: Advances in Neural Information Processing Systems 19 (NIPS 2006). No. 19. Cambridge, MA: MIT Press; 2007. pp. 545-552. Available from: https://resolver.caltech.edu/CaltechAUTHORS:20160315-111145907. ISBN: 0-262-19568-2

[14] Li Y, Hou X, Koch C, Rehg JM, Yuille AL. The secrets of salient object segmentation. In: IEEE Conference on Computer Vision and Pattern Recognition. 2014. pp. 280–287

[15] Xu K, Ba J, Kiros R, Cho K, Courville A, Salakhudinov R, et al. Show, attend and tell: Neural image caption generation with visual attention. In: International Conference on Machine Learning. 2015. pp. 2048–2057

[16] Itti L, Koch C, Niebur E. A model of saliency-based visual attention for rapid scene analysis. IEEE Transactions on Pattern Analysis and Machine Intelligence. 1998;**20**(11):1254-1259

[17] Borji A, Itti L. State-of-the-art in visual attention modeling. IEEE Transactions on Pattern Analysis and Machine Intelligence. 2013;**35**(1): 185-207. DOI: 10.1109/tpami.2012.89

[18] Bylinskii Z, DeGennaro EM, Rajalingham R, Ruda H, Zhang J, Tsotsos JK. Towards the quantitative evaluation of visual attention models. Vision Research. 2015;**116**:258-268. DOI: 10.1016/j.visres.2015.04.007

[19] Bruce NDB, Tsotsos JK. Saliency based on information maximization. In: Proceedings of the 18th International Conference on Neural Information Processing Systems. Cambridge, MA, USA: MIT Press. 2005. pp. 155–162

[20] Torralba A, Oliva A, Castelhano MS, Henderson JM. Contextual guidance of eye movements and attention in real-world scenes: The role of global features in object search. Psychological Review. 2006;**113**(4):766-786. DOI: 10.1037/ 0033-295x.113.4.766

[21] Borji A, Sihite DN, Itti L. What/ where to look next? Modeling top-down visual attention in complex interactive environments. IEEE Transactions on Systems, Man, and Cybernetics: Systems. 2014;**44**(5):523-538

[22] Han S, Vasconcelos N. Biologically plausible saliency mechanisms improve feedforward object recognition. Vision Research. 2010;**50**:2295-2307

[23] Borji A. Saliency prediction in the deep learning era: Successes and limitations. IEEE Transactions on Pattern Analysis and Machine Intelligence. 2019:669-700. DOI: 10.1109/TPAMI.2019.2935715

[24] Kümmerer M, Wallis TSA, Bethge M. DeepGaze II: Reading fixations from deep features trained on object recognition. ArXiv Preprint ArXiv: 1610.01563. 2016

[25] Pan J, Canton C, McGuinness K, O'Connor NE, Torres J, Sayrol E, et al. SalGAN: Visual saliency prediction with generative adversarial networks. In arXiv. 2017

[26] Russakovsky O, Deng J, Su H, Krause J, Satheesh S, Ma S. Imagenet large scale visual recognition challenge. International Journal of Computer Vision. 2015;**115**(3):211-252

[27] Qin Y, Lu H, Xu Y, Wang H. Saliency detection via cellular automata. In: Proceedings of the IEEE conference on computer vision and pattern recognition. IEEE Computer Society. 2015. pp. 110–119

[28] Li C, Yuan Y, Cai W, Xia Y, Dagan Feng D. Robust saliency detection via regularized random walks ranking. In: Proceedings of the IEEE Conference on Computer Vision and Pattern Recognition. 2015. pp. 2710-2717

[29] Zhao J, Sun S, Liu X, Sun J, Yang A. A novel biologically inspired visual saliency model. Cognitive Computation. 2014;**6**(4):841-848. DOI: 10.1007/ s12559-014-9266-z

[30] Yuan Y, Li C, Kim J, Cai W, Feng DDF. Reversion correction and regularized random walk ranking for saliency detection. IEEE Transactions on Image Processing. 2017;**1**:1-8. DOI: 10.1109/TIP.2017.2762422

[31] Wang W, Shen J. Deep visual attention prediction. IEEE Transactions on Image Processing. 2018;**27**(5): 2368-2378

[32] Hou X, Zhang L. Saliency detection: A spectral residual approach. In: 2007 IEEE Conference on Computer Vision and Pattern Recognition. 2007. pp. 1–8

[33] Zhu W, Liang S, Wei Y, Sun J. Saliency optimization from robust background detection. In: 2014 IEEE Conference on Computer Vision and Pattern Recognition. 2014. pp. 2814–2821

[34] Chen C, Tang H, Lyu Z, Liang H, Shang J, Serem M. Saliency modeling via outlier detection. Journal of Electronic Imaging. 2014;**23**(5):53023

[35] Tu Z, Ma Y, Li C, Tang J, Luo B. Edge-guided non-local fully convolutional network for salient object detection. 2019. Retrieved from: http://arxiv.org/abs/1908.02460

[36] Zhou Z, Wang Z, Lu H, Wang S, Sun M. Multi-type self-attention guided degraded saliency detection. Proceedings of the AAAI Conference on Artificial Intelligence, 34(07 SE-AAAI Technical Track: Vision). 2020. pp. 13082–13089. DOI: 10.1609/aaai.v34i07.7010

[37] Li G, Yu Y. Deep contrast learning for salient object detection. In: IEEE Conference on Computer Vision and Pattern Recognition. 2016. pp. 478–487

[38] Wang L, Lu H, Wang Y, Feng M, Wang D, Yin B, et al. Learning to detect salient objects with image-level supervision. In: 2017 IEEE Conference on Computer Vision and Pattern Recognition (CVPR). 2017. pp. 3796–3805. DOI: 10.1109/CVPR.2017.404

[39] Zeng Y, Zhuge Y, Lu H, Zhang L, Qian M, Yu Y. Multi-source weak supervision for saliency detection. 2019. Retrieved from: http://arxiv.org/abs/1904.00566

[40] He S, Jiao J, Zhang X, Han G, Lau RWH. Delving into salient object subitizing and detection. In: Proceedings of the IEEE International Conference on Computer Vision. 2017. pp. 1059–1067

[41] He K, Zhang X, Ren S, Sun J. Deep residual learning for image recognition. In: IEEE Conference on Computer Vision and Pattern Recognition. 2016. pp. 770–778

[42] Krizhevsky A, Sutskever I, Hinton GE. Imagenet classification with deep convolutional neural networks. In: Advances in Neural Information Processing Systems. 2012. pp. 1097–1105

[43] Deng J, Dong W, Socher R, Li L-J, Li K, Fei-Fei L. Imagenet: A large-scale hierarchical image database. In: IEEE Conference on Computer Vision and Pattern Recognition. 2009. pp. 248–255

[44] Selvaraju RR, Cogswell M, Das A, Vedantam R, Parikh D, Batra D. Grad-cam: Visual explanations from deep networks via gradient-based localization. In: ICCV. 2017. pp. 618–626

[45] Berga D, Fernández-Vidal XR, Otazu X, Pardo XM. SID4VAM: A benchmark dataset with synthetic images for visual attention modeling. ICCV. 2019: 8788-8797

[46] Glorot X, Bengio Y. Understanding the difficulty of training deep feedforward neural networks. In: Proceedings of the Thirteenth International Conference on Artificial Intelligence and Statistics. Proceedings of Machine Learning Research. 2010. Vol. 9. pp. 249-256. Available from: http://proceedings.mlr.press/v9/glorot10a.html

[47] Pan J, Cristian C, Kevin K, O'Connor NE, Torres J, Sayrol E, et al. SalGAN: Visual saliency prediction with generative adversarial networks. 2017

Chapter 5

# The Use of Leap Motion in Manual Dexterity Testing by the Box and Blocks Test: A Review Study

*Natália Regina Kintschner, Thiago Leandro Liporace, Silvana Maria Blascovi-Assis and Ana Grasielle Dionísio Corrêa*

## Abstract

This chapter aims to analyze works in the literature that show the therapeutic effects of using the Leap Motion Controller (LMC) sensor to assess hand fine motor dexterity, especially those involving the Box and Blocks Test. Besides the introduction, we will describe: (a) the LMC device and its forms of interaction in a Virtual Reality environment (immersive and non-immersive); (b) aspects of manual function assessment; (c) the functioning of the traditional Box and Blocks Test (BBT) and its virtual version (VBBT) developed with Virtual Reality technologies; (d) discussion about the VBBT integrated with the LMC, in physical therapy practice.

**Keywords:** moto rehabilitation, manual dexterity, leap motion, box and blocks test, virtual reality

## 1. Introduction

The Leap Motion Controller (LMC) is a small, portable, relatively low-cost device compared with other motion capture devices such as Doctor Kinetic®. It accurately detects all the hand and finger joints [1]. It is an interactive technology in Virtual Reality (VR) environments. LMC can be used in non-immersive mode, plugged into the computer via USB (desktop version), or attached to VR glasses (headset version) for immersive interaction. It has been currently being investigated as a technological resource to support upper limb motor rehabilitation interventions, as it allows capturing finer movements of the hands and fingers, which are essential for the rehabilitation of manual dysfunctions found in different conditions [2–6].

Several models of VR headsets are on the market, such as Gear VR, Rift glasses, and HTC Vive [2, 7–19]. Gear VR is a more cost-effective solution because it uses the screen of a Samsung smartphone as a viewing display [18]. The trend is that these technologies are available in people's homes, offering services in the most diverse areas of entertainment, education, and health.

Research with LMC points to its potential use by people with difficulties in fine and gross motor skills, explicitly concerning pincer grasp and power grip strength movements, extension and flexion of fists and fingers, forearm supination and



IntechOpen

pronation. Due to this extensive repertoire of gestures that LMC can detect, it is possible to find several studies with groups of people with stroke, older people with manual motor dysfunctions caused by aging [13], parkinsonians [8], children with developmental psychomotor disabilities, including cerebral palsy [20], Down syndrome [21], and autism [6].

The LMC allows measuring the motor performance of these people, such as reaction time, bimanual coordination, and the sequence of movements performed with the hands and fingers. For this reason, this remote sensing technology has shown promise for the rehabilitation field as it does not require the patient to wear motion detection devices (e.g., gloves with force and feedback sensors). Therefore, it provides a new interaction between the user and the computer, allowing for more natural and touch-free interaction. Hand dexterity in patients with upper limb motor disorders can be assessed from programmed tasks with graphic objects added to the virtual world [14].

Research involving the LMC integrated with the Box and Block Test (BBT) has emerged recently. Physical and occupational therapists use this test to verify manual function to assess and quantify the unilateral gross manual dexterity in children and adults. The BBT is made up of a wooden box with colored cubes, whose objective is to transport the cubes from one compartment of the box to the other in 1 minute. In the end, the number of cubes transferred per minute is counted. In its VR version, the LMC is used as an interaction device to transport blocks from one compartment to the other of the Box.

This chapter presents the works investigated using the LMC sensor integrated into the BBT. We are developing a version of BBT in VR that can be used on Desktop computers, both with video monitors and with VR glasses (HTC Vive). Then, we present details of this virtual BBT version in this chapter.

In addition to this introductory section, we divided this article into five sections: Section 2 provides details on how the LMC device works; Section 3 briefly presents the main tools for assessing manual function; Section 4 presents the functioning of the Box and Blocks Test and its virtual version in Virtual Reality developed by the researchers of our Game Therapy and Virtual Reality Laboratory; Section 5 shows scientific studies (performed with the Virtual BBT) found in the literature; Section 6 presents the conclusions.

The method adopted for the theoretical review was the query in indexed databases seeking information about the advantages and disadvantages of using the LMC device in immersive and non-immersive virtual reality situations. We sought to identify the errors and inaccuracies most commonly in the use of this device associated with the box and blocks test and the results found regarding the performance of the manual function.

## 2. Leap motion controller

Leap Motion Controller (LMC) is a small, portable device that accurately detects all the hand and finger joints [1]. It is a compact device, 8 cm wide by 3 cm high. The top of the device is made of smoked glass to hide the two image sensors and infrared LEDs that work together to track the user's hand movements (**Figure 1**).

It is possible to use the LMC connected directly to the computer in non-immersive experiences (**Figure 2a**). In this case, the video monitor is used as a viewing device. In immersive experiences, the LMC is coupled to VR glasses, such as a Gear VR. (**Figure 2b**). The simplicity of the LMC could facilitate the approach to

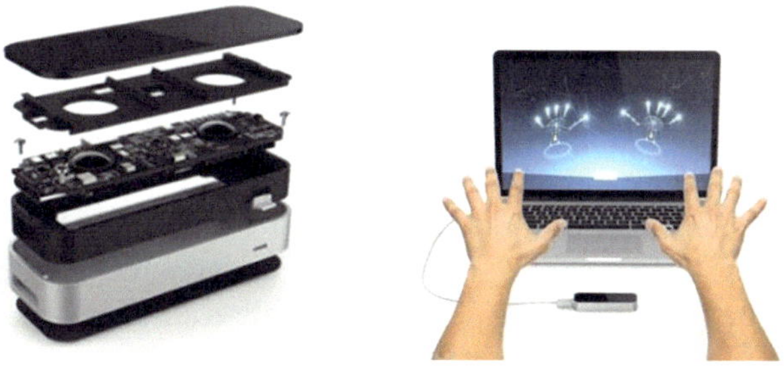

**Figure 1.**
*Leap motion controller (LMC). Source: Leap motion, 2019 https://www.leapmotion.com/.*

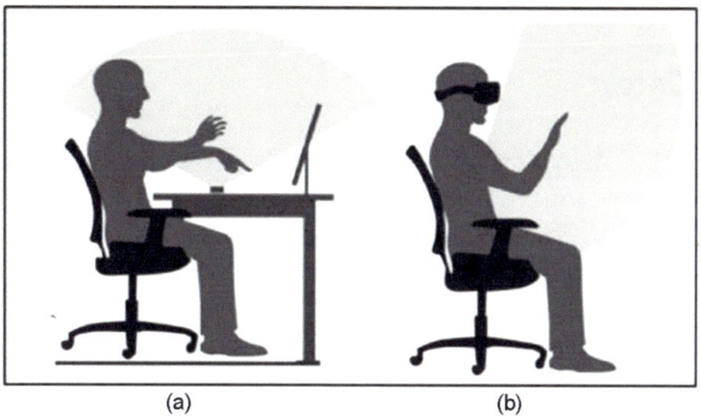


**Figure 2.**
*(a) LMC with non-immersive interaction (desktop). (b) LMC with immersive interaction (VR headset). Source: Leap Motion, 2019 https://www.leapmotion.com/.*

technology, increasing the feeling of immersion, imagination, and interaction with the virtual environment [4].

The LMC works with accuracy down to 1/100 mm without visible latency in its visual field. The viewing range is 80 cm above and around the device. This limit occurs due to the propagation of the LED light through space since it is difficult to infer the hand's position from a certain distance.

The LMC is postulated as a playful tool that can favor the inclusion of participants in a different environment, in which they can face new challenges and achieve new goals by interacting in real time with hand and finger movements, functioning as an active therapy that requires high commitment and motivation [9].

In a study by [9], the authors conclude that the LMC is the primary haptic VR sensor for upper limb mobility recovery compared with other non-immersive VR devices, such as the Doctor Kinetic® and Nintendo® Wii game systems, which are more specialized in posture and balance. The results suggest that the LMC can be considered a valuable and effective haptic VR device to improve different aspects of upper limb motor function in neurological patients [9].

## 3. Manual function assessment

The functional assessment of the hands is part of the therapeutic routine of several professionals who work with strategies for functional recovery of the upper limbs. For this, several tests evaluate the precision and speed of movements, proposing different tasks in standardized tests and validated in other countries. Reference [22] analyzes, through a critical review, several tools for assessing manual dexterity, referring to their psychometric properties. Among the tools reviewed are tests commonly used for upper limb assessment, such as the Minnesota Dexterity Test, the Nine Hole Peg Test (9HPT), the Jebsen-Taylor Hand Function Test (JTHFT), and the Box and Blocks Test (BBT).

These conventional tests for assessing manual function have several limitations, such as the long time to be done, the need for a highly trained professional to determine the patient's outcome, the patient's displacement, and the reliability of the performance evaluation, which is low [23]. The use of technology associated with these tests is a way of fulfilling these principles since it can provide exercises in a controlled, repetitive, intensive, interactive, and motivating way [22].

## 4. Box and blocks test

A. Jean Ayres and Patricia Holser Buehler created the first Box and Blocks Test (BBT). They used a bowl and blocks to assess gross manual dexterity in adults with cerebral palsy [24]. Later, Patricia Holser Buehler and Elizabeth Fuchs changed the shape of the test to a gift box, obtaining copyright in 1957 [16].

As described by [16, 17], the BBT (**Figure 3**) is used to verify manual function to assess and quantify unilateral gross manual dexterity in children and adults. It consists of a wooden box, 53.7 cm long, with a partition, also made of wood, higher than the edges of the box, separating it into two compartments of equal dimensions. The blocks, also made of wood and in the form of colored cubes (primary colors), are 150 in number, measuring 2.5 cm on a side divided equally by color. The box's long sides are 53.7 cm by 8.5 cm and are nailed 1 cm thick from the base. The short ends are 7.5 cm by 25.4 cm and have been fixed to the top of the bottom between the long sides.

As a prerequisite for the application of the test, a quiet environment is required, with the examinee seated in a chair suitable for his/her height. The wooden box should be placed horizontally in front of him so he can fully view the area and equipment in question. It consists of moving as many cubes as possible from one compartment to another for 1 minute [25]. The BBT was proposed [16] with parameters between 20 and 94 years old. In Brazil, the BBT was used for the age group between

**Figure 3.**
*Box and blocks test (BBT). Source: MENDES et al., 2001 [17].*

15 and 86 years, with typical groups and multiple sclerosis [17]. For the lower ranges, from 7 to 14 years old, the test parameters for Brazilians were mentioned by [12].

These functional measures of dexterity contrast with neuroscientific investigations that show that skill is multicomponent, including the ability to control force, control the timing of movements, execute independent finger movements, and execute motor sequence [26].

### 4.1 The virtual box and blocks test

Our researchers from the Game Therapy and Virtual Reality Laboratory (Lab GameVR) created a virtual application of BBT, which we call Virtual BBT (desktop version and HTC Vive). The main requirements raised were the following:

- Availability of two forms of use:
- Conventional: with the countdown starting at 60 seconds.
- Training: with progressive time without time counting.
- On-screen display of the score to the user.
- Storage of collected data.
- Help feature with usage instructions.
- Bilateral hand training.

Based on these requirements, a game prototype was conceived in VR using the Unity 3D game engine integrated into the LMC. When executing the game file, the Virtual BBT start screen is presented to the user with the following configuration and selection options (**Figure 4**) "Right or left hand," "Start," "Training," and "? [Help]." At that moment, the user must plug the LMC into the computer where the application is running.

**Figure 4.**
*Virtual BBT home screen. Source: Author.*

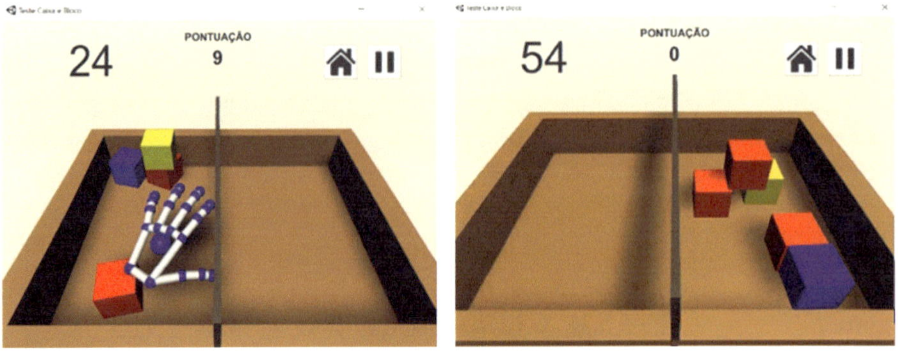


**Figure 5.**
*Virtual BBT screen in run mode. Source: Author.*

When selecting the "Start" option, the Virtual BBT screen is shown to the user (**Figure 5**). The screen shows the Wooden Box and the blocks modeled with Unity 3D. The user can observe the virtual hand moving around the scenario by placing the hands on the LMC. The screen displays the timer counting down. In this game scenario, the user has 60 seconds to transfer the blocks from one box compartment to the other. The same game scenario is displayed by selecting the "Training" option but without the countdown. In this scenario, the user can stay as long as he wants to interact with the Virtual BBT. An immersive version of BBT Virtual is being developed for use with HTC Vive.

At the end of use, the data referring to the player's score, date, time, patient's name, time, and laterality score of the executed hand are stored locally on the user's computer in a file in the csv format.

## 5. Discussion

Scientific studies involving box and blocks test in its virtual version are emerging in the context of rehabilitation in patients with neurological diseases, such as Parkinson's Disease (PD), Cerebral Vascular Accident (CVA), and Multiple Sclerosis (MS), among others, intending to study the validity, feasibility, and psychometric properties of the test [3, 10, 15]. However, few studies have developed the virtual BBT (VBBT) [10].

The interest in developing virtual tests is multiple [10]. It can reduce the inter-observer subjectivity in the classic assessment, providing a complete virtual rehabilitation at home where patients can assess their cognitive and motor improvements using validated virtual evaluation.

**Table 1** shows some studies that addressed the topic, including author, study population, research country, goals, method, and the main results.

Researchers [3] assessed the validity of BBT and VBBT in participants with Parkinson's Disease (PD). They developed the VBBT using an immersive headset (Rift eyewear) and the LMC to evaluate unilateral manual dexterity in this audience. Participants were instructed to perform the physical BBT (once) and the virtual BBT (twice, one immersive and one non-immersive) separately. The results indicated a moderate correlation between the physical BBT and the VBBT scores.

| Author | Study population | Research country | Goals and method | Main results |
|---|---|---|---|---|
| Onã et al., 2020. | Participants with Parkinson's Disease (PD) | Spain | Developed the VBBT using an immersive headset and the LMC to evaluate unilateral manual dexterity. Participants performed the physical BBT (once) and the virtual BBT (twice, one immersive and one non-immersive) separately. | 3D depth perception allowed the movement of more cubes in the immersive virtual BBT regarding BBT. |
| Giersen et al., 2016. | Typical Individuals | United States | Present their version of the VBBT and the results of a pilot study in which participants completed the BBT and VBBT, comparing their scores and opinions. | The number of video games and VR experience was positively correlated with task performance. |
| Alvarez-Rodriguez et al., 2020. | Neurological Diseases | Spain | Developed a form of application of the VBBT using the LMC. The sample consisted of 24 individuals, divided into two groups: the typical group (n = 12) and the group with neurological diseases (n = 12). The study is conducted in a single experimental session by performing the physical and virtual BBT with the dominant hand. | BBT and VBBT high-performance showed corresponding final results, with a high tendency between the two tests. |
| Teruel et al., 2019. | Patients who suffer a spinal cord injury at the cervical level and affects the function of the upper limb. | Spain | Present a tool developed from the BBT for its virtual version, using the LMC and Unity 3d to supervise the therapy execution. | VBBT enables the therapists to customize therapy according to each patient's specific needs. |

*Source: Author.*

**Table 1.**
*Summary of studies.*

The results show that the 3D depth perception allowed the movement of more cubes in the immersive virtual BBT regarding BBT. This study presented three relevant findings: the gap in the number of blocks transported in the physical and virtual systems seems to tend to be a constant, the correlation between the obtained result between the physical and virtual systems is statistically significant, and the test-retest analysis shows an excellent statistically considerable correlation between attempts with the virtual system. In this sense, the relationship between physical and virtual systems can be improved using fully immersive VR due to better depth perception.

In a study [11], researchers present their version of the VBBT and the results of a pilot study in which participants completed the BBT and VBBT, comparing their scores and opinions. The authors also compared how the participants handled time during both versions running. The Unity 3d platform and the LMC device were used for the virtual version. For gameplay, timers were implemented for the participants' 15-second practice and the 60 seconds of the entire session. During the whole session, the LMC data were recorded for future analysis. These data include participant position, palm position, fingertip position, and joint angles.

The results compared the scores and opinions of participants who took both versions of the test, showing that the number of video games and VR experience was positively correlated with task performance. The results also showed that participants with less knowledge of video games and VR performed in a slower time than those with more experience, who reported greater enjoyment than those who did not have VR contact. Finally, the authors conclude that the participants found the virtual version more frustrating. However, they still preferred to deal with this version rather than the physical one to have one more chance at performance. On average, healthy participants moved more 35 blocks in the BBT than in the VBBT.

In another study by [27], they developed a form of application of the VBBT using the LMC. The sample in the research consisted of 24 individuals, divided into two groups: the typical group (n = 12) and the group with neurological diseases (n = 12) and, consequently, impairment of upper limb motor function. The study is conducted in a single experimental session by performing the physical and virtual BBT with the dominant hand. Firstly, the physical BBT was performed, starting with a practice test lasting 15 seconds. Once the test was completed, the participants had a 5-minute rest before running the VBBT. Before beginning this version, there was also a practical period in which the participant performed two 1-minute tests to make him comfortable and familiar with the virtual environment. The 15-second test period started with the dominant or less affected hand, followed by 60 seconds with the automatic counting of correctly transported blocks.

This research presented some limitations reported by the authors, such as difficulties with the technology used (LMC), whose performance was strongly conditioned by the environmental conditions and the computer's performance. However, this study showed that considering the physical BBT, upper limb motor skill was statistically higher in the typical population than in the people with neurological diseases. The same behavior was observed for the VBBT. Besides, within each analyzed group, the performance in the BBT was superior to that of the VBBT. However, BBT and VBBT high performance showed corresponding final results, with a high tendency between the two tests.

The VBBT version developed in this work showed high consistency in its application in a sample of typical individuals and patients with neurological diseases. According to the authors, the next step is that the VBBT, integrated with LMC, serves as an element to assess motor dysfunctions, allowing manual dexterity training in the population with neurological diseases.

Other researchers [28] present a tool developed from the BBT for its virtual version, using the LMC and Unity 3d. The authors included some feedback in this version

(VBBT) to increase the patients' motivation, with the intention that they become aware of their competence while performing exercises for rehabilitation. The developed game can be configured according to the needs of each one. It offers facilities to define the dominant hand, the number of blocks the patients will have to pick up and their size. After that, the game starts so that the player can interact with the blocks, lasting until all the blocks have been placed on the non-dominant side area of the screen or until the participant gives up playing.

During the VBBT execution, patients receive information about the current game and their progression concerning previous moves. While the participant performs the task, the game simultaneously shows the number of moved blocks (score), the number of blocks they need to move to finish the game correctly (goals), the elapsed time, and the current speed related to the number of blocks moved per minute. In addition, to complete this information, a dynamic status bar is also shown to describe the relationship between the patient's current speed and the speed of their previous best and worst performance. The value of this status bar changes in real time, decreasing if the patient does not move new blocks or increasing when a new block is moved. When finished, additional information is presented to the patient on a result screen, showing the player's final score, goal, elapsed time, and speed.

The authors emphasize that upper limb rehabilitation has become a critical need due to the number of people affected by this condition when they have a neurological disease, for example. They also point out that the existing treatments for these conditions are demanding and expensive. Therefore, the VBBT enables the therapists to customize therapy according to each patient's specific needs, exploring an important feature, which is the introduction of different motivation facilities to attract this audience to perform a repetitive task, which, on the other hand, could be tedious.

From the studies discussed, it can be noticed some of the advantages and disadvantages of using the LMC [27], which can be a device dependent on the conditions presented by the computer to which it will be connected. Some results [11] showed that participants with less knowledge of video games and VR performed in a slower time than those with more experience, who reported greater enjoyment than those who did not have VR contact. Therefore, the use of LMC may also be related to the individual's previous contact with technology.

When used with the BBT, the LMC proved to be more acceptable in its immersive version [3], mainly because users are able to transfer more blocks when inserted in the 3d view, which consequently generates less fatigue in the manual function and an increase in the motivation. Another advantage of using this technology is that together with the virtual BBT, health professionals have the possibility to meet the needs of each patient individually, through time, the number of blocks, feedbacks, among others [28].

It is important consider that the use of MC associated with the virtual BBT allows the execution of the task of transferring blocks in a situation of evaluation and training for the manual function without the need to the physical blocks, reducing the chance of any type of contamination, BBT the manual dexterity, promoting the development and health of users.

Therefore, there is a need for expansion in studies that encompass the rehabilitation of manual dexterity, technology, and its devices and the tests involved in this context, such as BBT in its virtual version.

## 6. Conclusions

This chapter aimed to present an overview of a research development using the Virtual BBT. It is noteworthy that the development of BBT, in its immersive and non-immersive virtual version integrated with the LCM device, presents itself as a possibility for testing and assessing patients with or without changes in manual dexterity. This technology emerges as an innovative and motivating way for use in various areas, such as health. Then, more research is needed with different audiences to investigate its usability and effectiveness of both (virtual BBT and the LMC device) and their forms of use so that technical issues can be increasingly improved in the use of these equipment.

## Conflict of interest

The authors declare no conflict of interest.

This study was financed in part by the Coordenação de Aperfeiçoamento de Pessoal de Nível Superior – Brazil (CAPES).

## Author details

Natália Regina Kintschner, Thiago Leandro Liporace, Silvana Maria Blascovi-Assis* and Ana Grasielle Dionísio Corrêa
Game Therapy and Virtual Reality Laboratory (Lab GameVR) Postgraduate Program in Developmental Disorders (Programa de Pós-Graduação em Distúrbios do Desenvolvimento [PPGDD]), Mackenzie Presbyterian University, São Paulo, Brazil

*Address all correspondence to: silvanam.assis@mackenzie.br

## References

[1] Afyouni I, Qamar AM, Hussain SO, Rehman FU, Sadiq B, Murad A. Motion-based serious games for hand assistive rehabilitation. In: Proceedings of the 22nd International Conference on Intelligent User Interfaces Companion (IUI '17 Companion). New York, NY, USA: Association for Computing Machinery; 2017. pp. 133-136

[2] Niechwiej-Szwedo E, Gonzalez D, Nouredanesh M, Tung J. Evaluation of the Leap Motion Controller during the performance of visually-guided upper limb movements. PLoS One. 2018;**13**(3):e0193639

[3] Oña ED, Balaguer C, Cano-de la Cuerda R, Collado-Vázquez S, Jardón A. Effectiveness of serious games for leap motion on the functionality of the upper limb in Parkinson's Disease: A feasibility study. Computational Intelligence and Neuroscience. 2018;**2018**:7148427

[4] Tarakci E, Arman N, Tarakci D, Kasapcopur O. Leap Motion Controller-based training for upper extremity rehabilitation in children and adolescents with physical disabilities: A randomized controlled trial. The Journal of Hand Therapy. 2020;**33**(2):220-228

[5] Wu YT, Chen KH, Ban SL, Tung KY, Chen LR. Evaluation of leap motion control for hand rehabilitation in burn patients: An experience in the dust explosion disaster in Formosa Fun Coast. Burns. 2018;**45**(1):157-164. DOI: 10.1016/j.burns.2018.08.001

[6] Zhu G, Cai S, Ma Y, Liu E. A series of leap motion-based matching games for enhancing the fine motor skills of children with autismo. In: IEEE 15th international conference on advanced learning technologies. Vol. 2015. 2015. pp. 430-431

[7] Borrego A, Latorre J, Alcañiz M, Llorens R. Comparison of oculus rift and HTC Vive: Feasibility for virtual reality-based exploration, navigation, Exergaming, and rehabilitation. Games for Health Journal. 2018;**7**(3):151-156

[8] Butt AH, Rovini E, Dolciotti C, Bongioanni P, De Petris G, Cavallo F. Leap motion evaluation for assessment of upper limb motor skills in Parkinson's disease. IEEE International Conference on Rehabilitation Robotics. 2017;**2017**:116-121

[9] Cortés-Pérez I, Zagalaz-Anula N, Montoro-Cárdenas D, Lomas-Vega R, Obrero-Gaitán E, Osuna-Pérez MC. Leap motion controller video game-based therapy for upper extremity motor recovery in patients with central nervous system diseases: A systematic review with meta-analysis. Sensors (Basel). 2021;**21**(6):2065

[10] Everard G, Otmane-Tolba Y, Rosselli Z, Pellissier T, Ajana K, Dehem S, et al. Concurrent validity of an immersive virtual reality version of the box and block test to assess manual dexterity among patients with stroke. Journal of NeuroEngineering and Rehabilitation. 2022;**19**(1):7. DOI: 10.1186/s12984-022-00981-0

[11] Gieser SN, Gentry C, LePage J, Makedon F. Comparing objective and subjective metrics between physical and virtual tasks. In: Lackey S, Shumaker R, editors. Virtual, Augmented and Mixed Reality. Cham: VAMR, Springer; 2016. p. 9740

[12] Guimarães R, Blascovi-Assis S. Uso do teste caixa e blocos na avaliação de destreza manual em crianças e jovens com síndrome de Down. Revista de Terapia

Ocupacional da Universidade de São Paulo; Brasil. 2012;**23**(1):98-106. DOI: 10.11606/issn.2238-6149.v23i1p98-106

[13] Iosa M, Morone G, Fusco A, et al. Leap motion controlled videogame-based therapy for rehabilitation of elderly patients with subacute stroke: A feasibility pilot study. Topics in Stroke Rehabilitation. 2015;**22**(4):306-316

[14] Kintschner NR, Correa AGD, Blascovi-Assis SM. Realidade virtual controlada por sensores de detecção de movimentos das mãos aplicada aos transtornos do desenvolvimento. In: Cibelle Albuquerque de la Higuera Amato; Decio Brunoni; Paulo Sérgio Boggio. (Org.). Distúrbios do Desenvolvimento: Estudos Interdisciplinares. 1 ed. São Paulo: Editora Memnon, pp. 443-454

[15] Martínez-Piédrola RM, García-Bravo C, Huertas-Hoyas E, et al. The influence of self-perception on manipulative dexterity in adults with multiple sclerosis. Occupational Therapy International. 2021;**2021**:5583063

[16] Mathiowetz V, Volland G, Kashman N, Weber K. Adult norms for the box and block test of manual dexterity. The American Journal of Occupational Therapy. 1985;**39**(6):386-391

[17] Mendes M, Tilbery C, Balsimelli S, Moreira M, Barão C. Teste de destreza manual da caixa e blocos em indivíduos normais e em pacientes com esclerose múltipla. Arquivos de Neuro-Psiquiatria. 2001;**59**(4):889-894

[18] Moro C, Štromberga Z, Stirling A. Virtualisation devices for student learning: Comparison between desktop-based (Oculus Rift) and mobile-based (Gear VR) virtual reality in medical and health science education. Australasian Journal of Educational Technology. 2017;**33**(6):1-10

[19] Ogdon DC. HoloLens and VIVE pro: Virtual reality headsets. Journal of the Medical Library Association: JMLA. 2022;**107**(1):118

[20] Oliveira HB, Campos FW, JMT C. A study applied to the validation of the box and blocks manual dexterity virtual test with non-disabled users. In: 21st Symposium on Virtual and Augmented Reality (SVR), Rio de Janeiro, Brazil. 2019. pp. 206-215. DOI: 10.1109/SVR.2019.00045

[21] Sanchez V, Cruz O, Solorza E, Encinas I, Caro K, Castro L. BeeSmart: A gesture-based videogame to support literacy and eye-hand coordination of children with down syndrome. In: International Conference on Games and Learning Alliance. Computer Vision and Image Understanding. Vol. 1. Cham, Switzerland: Springer; 2017. pp. 43-53. DOI: 10.1007/978-3-319-71940-5_4

[22] Yancosek KE, Howell D. A narrative review of dexterity assessments. Journal of Hand Therapy. 2009;**22**(3):258-270

[23] Schallert W, Fluet MC, Kesselring J, Kool J. Evaluation of upper limb function with digitizing tablet-based tests: Reliability and discriminative validity in healthy persons and patients with neurological disorders. Disability and Rehabilitation. 2022;**44**(8):1465-1473

[24] Silva GL, Ceron BM, Borba KM, Amaral DS, de Queiroz Marcelino JF, de Sales MD, et al. Repercussões do treinamento com realidade virtual não imersiva nas habilidades motoras manuais de pessoas com doença de Parkinson. Acta Fisiátr. [Internet]. 2019;**26**(1):43-48

[25] Soares N, Pereira G, Italiano R, Morais G, Melo S. Terapia baseada em realidade virtual usando o Leap Motion Controller para reabilitação do membro superior após acidente vascular cerebral. Scientia Medica. 2017;**27**:25935

[26] Rabah A, Le Boterff Q, Carment L. A novel tablet-based application for assessment of manual dexterity and its components: A reliability and validity study in healthy subjects. Journal of NeuroEngineering and Rehabilitation. 2022;**19**(35)

[27] Alvarez-Rodríguez M, López-Dolado E, Salas-Monedero M, Lozano-Berrio V, Ceruelo-Abajo S, Gil-Agudo A. de los Reyes-Guzmán a. concurrent validity of a virtual version of box and block test for patients with neurological disorders. World. Journal of Neuroscience. 2020;**10**:79-89

[28] Teruel M, de los Reyes-Guzmán A, Villanueva J, Lozano-Berrio V, Alvarez-Rodríguez M, Ceruelo-Abajo S, et al. Picking Cubes: A Rehabilitation Tool for Improving the Rehabilitation of Gross Manual Dexterity. Vol. 806. Cham: Springer; 2019. pp. 265-273. DOI: 1007/978-3-030-01746-0_31

Chapter 6

# Precise 6DOF Localization of Robot End Effectors Using 3D Vision and Registration without Referencing Targets

*Liang-Chia Chen, Sheng-Hao Huang and Bo-Han Huang*

## Abstract

A method for detecting the precise 6-degree-of-freedom (6DOF) localization of robotic arms end effectors without referencing any additional feature target during in-line robot operation is introduced to facilitate precise robot positioning and monitoring. In this work, a 3D vision probe with digital structured-light projection is integrated with a wafer handling robot to perform online 6DOF location monitoring in semiconductor production. Precise alignment of the robotic arms end effector moving in the 3D operation space is realized by robust point cloud object alignment using regional surface area descriptors and the variant iterative closest point algorithm. Verified and confirmed by experimental tests, the developed method can achieve online 6DOF location monitoring with micron-level accuracy. Moreover, the proposed method can completely avoid the disadvantages of existing methods, namely relying on planar 2D images and demanding an additional target to be embedded with the end effector for localization, which reduces practical application, especially in an in-line operation environment. The major technical breakthrough of the present work is the target-free precise 6DOF localization of moving objects.

**Keywords:** 6DOF localization, iterative closest point, point cloud alignment, 3D vision, robot end effectors

## 1. Introduction

Robotic arms are usually used in automated production lines to perform repetitive or hazardous operations to ensure high manufacturing quality. For instance, a wafer handling robot is usually used in the automated production line of semiconductor components for sending or taking wafers between cassettes at etching, deposition, and photolithography stations with demanding accuracy, repeatability, and reliability [1]. However, potential induced stress or deformation during continuous robotic arm operation may cause the end effector position and orientation to deviate from the position specified, which would result in the collapse of the processing unit with the

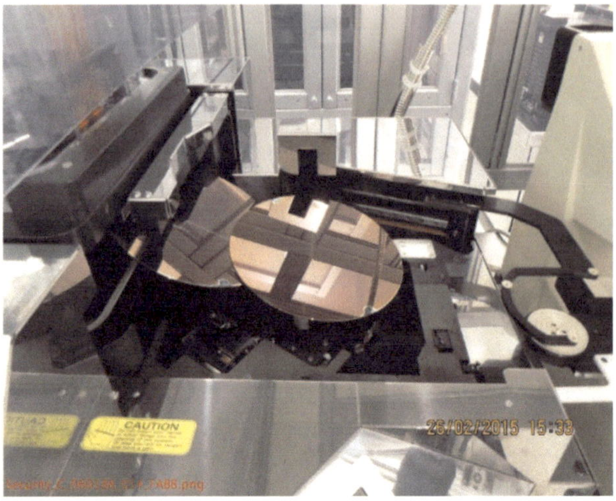


**Figure 1.**
*Wafer damage caused by positioning deviation of robot end effectors during manufacturing.*

wafer scrapped or crushed, as shown in **Figure 1**, thus incurring significant production losses. To avoid such an undesired scenario, this study proposes a robust and precise monitoring method for performing the 6-degree-of-freedom (6DOF) localization of robotic arms end effectors.

Recent studies have investigated the problems related to position uncertainties of robotic arms end effectors and analyzed the life cycle of the robotic arm to improve the reliability of the wafer handling robot [2, 3]. However, in general, planar translation of end effectors has been measured using only 2D imaging; with depth information missing, 6DOF localization of end effectors cannot be realized.

To obtain accurate variations in both position and orientation of the robotic arms end effector, this study developed a 3D machine vision probe with point clouds measured for achieving 6DOF localization of the end effector. Point cloud registration is crucial to determining precise object 6DOF transformation between different locations in the 3D space and involves the coarse alignment stage and the refined model alignment stage. During coarse alignment, the object point cloud aligns only approximately with the target (reference) point cloud. Methods used include spin image [4], point signature [5], and regional surface area descriptor [6], with different levels of accuracy, efficiency, and robustness achieved. This initial matching between the object and the target model is then followed by fine model alignment for more precise matching of the two. The most widely adopted algorithm is the iterative closest point (ICP) [7], which uses singular value decomposition (SVD) to find the set of closest point clouds between the object and the target model. It refines the deviation between the corresponding points until the least-squares error is minimized.

This chapter presents a novel 6DOF detection method that uses a developed structured-light 3D scanner [8] to obtain the 3D information of the robotic arms end effector and then performs point cloud alignment to detect the 6DOF variations of the robotic arms end effector. The process of the proposed method is illustrated in **Figure 2** and described in detail in Section 2.

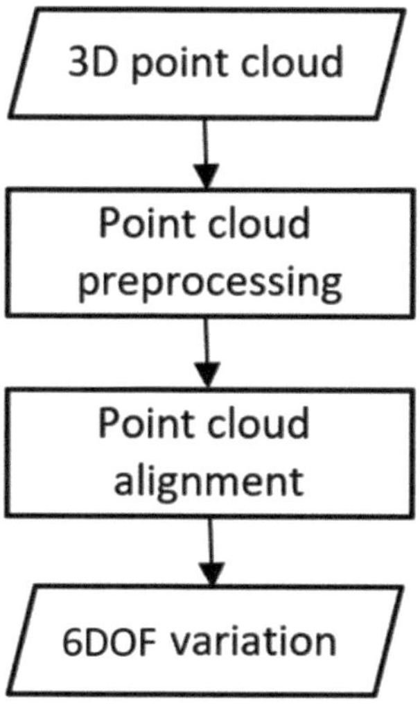


**Figure 2.**
*Process of the proposed method.*

## 2. Methods

### 2.1 Point cloud preprocessing

#### *2.1.1 Denoise algorithm*

Optical measurement is susceptible to noise and outliers caused by the external environment, object surface material, and reflectivity, and these noises and outliers must be eliminated before point cloud registration. In this study, Statistical Outlier Removal (SOR) [9] was applied to neighboring points of each point in the point cloud cluster. First, the average distance of each point to their K nearest neighbors is calculated. The average distance $d_{mean}$ and standard deviation σ are then used as the threshold for determining noise or outlier. The distance threshold can be expressed as:

$$d_{thresh} = d_{mean} + K * \sigma \tag{1}$$

where K is a scale factor and is usually set to equal 1.0. **Figure 3** illustrates the outlier determination and removal. First, the distances of the nearest K points $P_j$ (j = 0,1,2,3, ... ,k) of the point $P_i$ are calculated. The average distance between $P_o$ and its neighboring point $P_j$ is $d_i$; if $d_i$ is greater than the distance threshold, $d_{thresh}$, Po is regarded as an outlier and is thus removed from the point set P.

**Figure 4a** shows the original point cloud, while **Figure 4b** shows the point cloud with the outlier removed with a multiplicity factor of 1.0 after searching the neighboring points.

#### *2.1.2 Downsampling strategy*

The structured-light measurement probe uses a high-resolution sensor of up to two megapixels; hence, the amount of data for the reconstructed point cloud is significant and dense, often reaching hundreds of thousands of points. It is thus necessary to

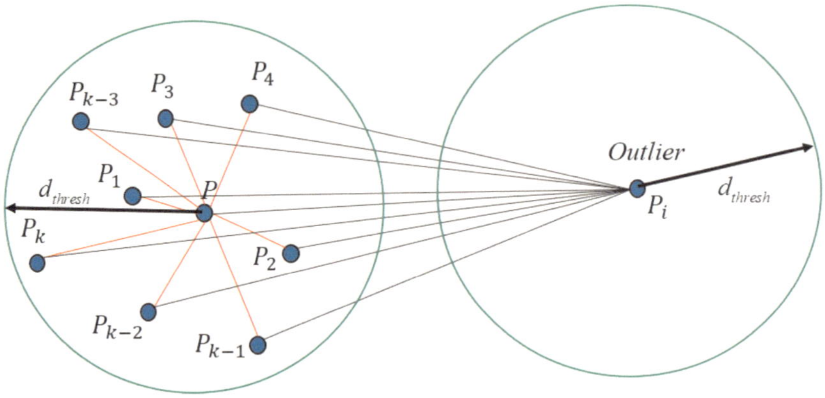


**Figure 3.**
*Illustration of outlier removal.*

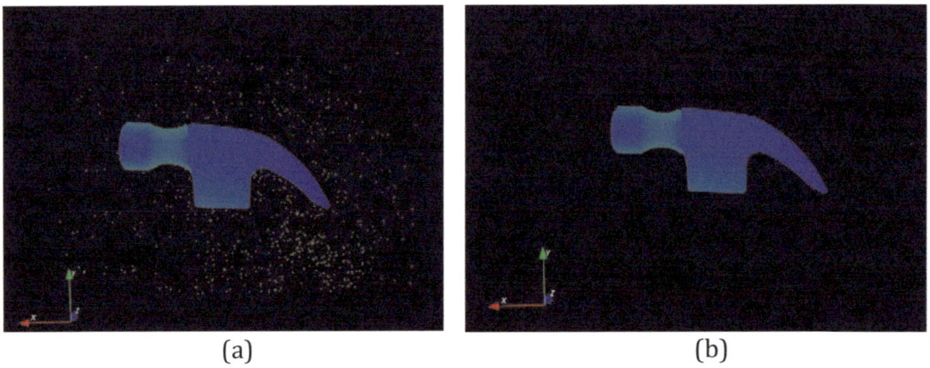


**Figure 4.**
*(a) Original point cloud and (b) point cloud after outlier removal.*

effectively reduce the number of data points while strictly preserving the morphological characteristics of the object for evaluating pose variation.

The widely used point-cloud downsampling algorithms are voxel grid filtering and Delaunay Triangulation. Pixel is the basic unit that constitutes an image, and voxel means a cubic grid, which is the unit cube that constitutes a 3D space. For data reduction, the voxel grid filter replaces all the points in the voxel with the mass center of all the points. All the points in the voxel are represented by the n points belonging to the unit voxel, as shown in **Figure 5**.

Delaunay triangulation [10] allows all input points to form a convex polygon (convex hull) [11], as shown in **Figure 6**, and all triangles within the convex polygon must satisfy the property of maximizing the minimum interior angle and the noncircularity of any four points. If Delaunay triangulation is applied to the 3D space, the four adjacent points can form a tetrahedron and create an external sphere, the so-called Alpha-ball, as shown in **Figure 7**. The red circle denotes the external sphere of the tetrahedron, and the excess points can be removed using the Alpha-ball algorithm [12], thus achieving data reduction while preserving the object's surface characteristics.

As shown in **Figure 8a**, a precision ceramic gauge block with sharp edges was used as the test object to compare the above two-point cloud downsampling methods. The

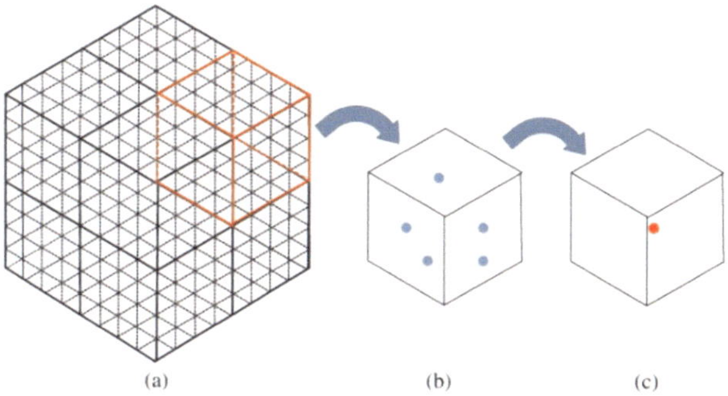


**Figure 5.**
*Illustration of voxel-grid filtering: (a) unfiltered voxel; (b) unfiltered unit voxel; and (c) the mass center of the unit voxel.*

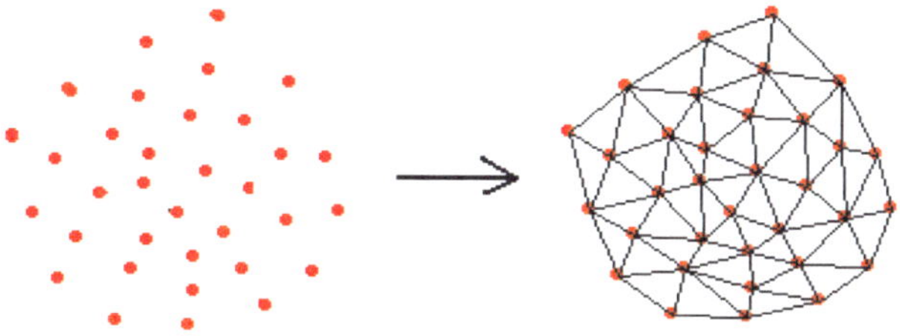

**Figure 6.**
*Illustration of convex hull formed by Delaunay triangulation.*

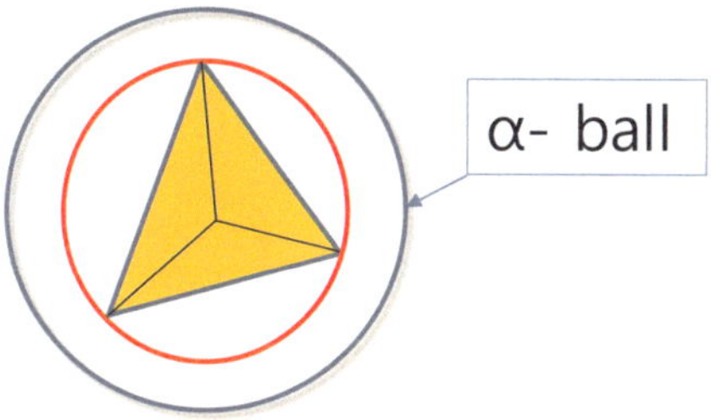


**Figure 7.**
*Illustration of alpha-ball [12] formed by applying Delaunay triangulation to 3D space.*

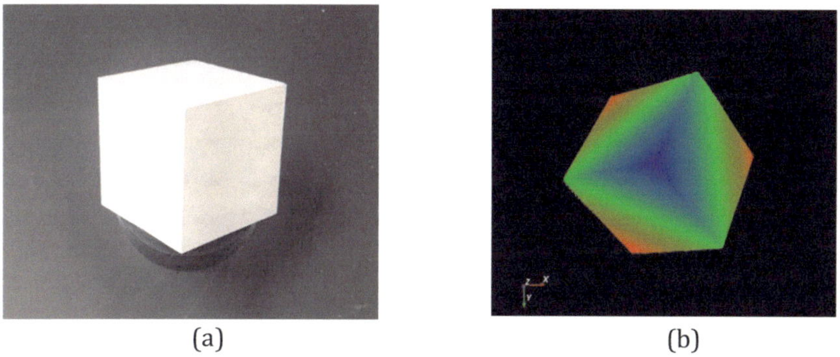


**Figure 8.**
*(a) Original and (b) reconstructed gauge block after outlier removal.*

reconstructed result after outlier removal is shown in **Figure 8b**, and the number of points is enormous, totaling 300,495.

Such a large amount of data typically requires a long computation time. Therefore, voxel grid filtering and Delaunay triangulation are performed in the proposed method to reduce the amount of 3D reconstructed point cloud data. Their performances are compared in terms of the integrity of object features retained. **Figure 9a** shows the result of the original point cloud data after voxel grid filtering; the total number of points becomes 9239, reduced by 96.9%; and **Figure 9b** shows the registration result with an average registration error of 40.8 μm between the original and downsampled point cloud, and the maximum error of 520 μm at the object's corner edge. **Figure 9a** shows the result of the original point cloud data after Delaunay triangulation; the total number of points becomes 9253, reduced by 96.9%; and the registration result, shown in **Figure 10b**, has an average registration error of 10 μm and the maximum error of 90 μm.

Comparing the two-point cloud downsampling methods revealed that Delaunay triangulation can achieve a much smaller registration error than voxel grid filtering, especially in the edge area of the measured object. Voxel grid filtering uses average filtering, which smooths out the edge features of the object. In contrast, Delaunay triangulation removes only unnecessary points, thus preserving the edge features of the object. Hence, using Delaunay triangulation to reduce the number of data points as inputs for the subsequent algorithm can obtain better accuracy when calculating pose variation and avoid the error generated by the preprocessing of point clouds.

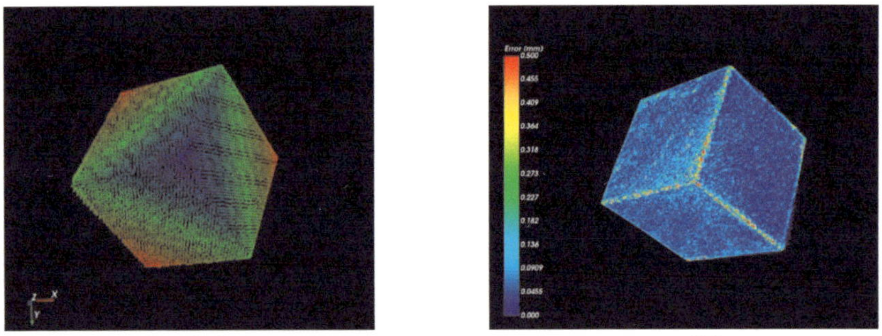

(a) (b)

**Figure 9.**
*(a) Original point cloud after voxel grid filtering and (b) registration result.*

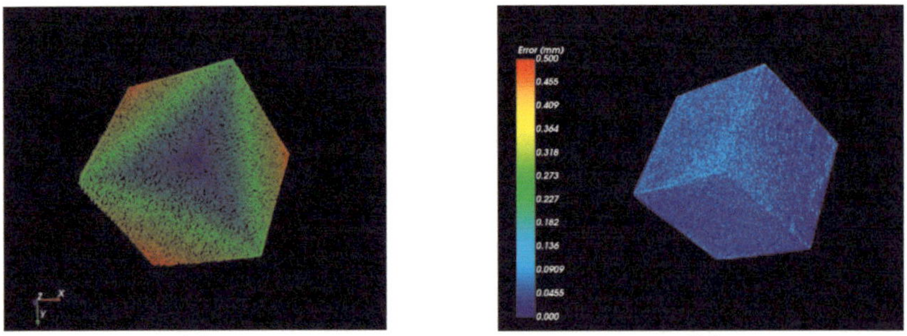


(a) (b)

**Figure 10.**
*(a) Original point cloud after Delaunay triangulation and (b) registration result.*

## 2.2 Coarse alignment between scanned data and model object

This study used the regional surface area descriptor (RSAD), proposed by Chen et al. [6], to determine the geometric relation between 6DOF of two sets of point clouds measured on an object of two different locations and orientations. The method assumes that the same surface area distribution defined in the oriented bounding box (OBB) [13] of a point cloud set can be employed to detect the 6DOF of the object in 3D space. Thus, to find the most matched pose, the RSAD of the measured point cloud is iteratively compared with the feature descriptor generated by the 3D virtual camera from the model point cloud, which represents the target model or the original pose of the object. **Figure 11** illustrates the matching method using the RSAD.

### *2.2.1 Database generation*

First, a coordinate system is established at the center of the model point cloud, and the virtual camera is located on the z-axis of the model point cloud coordinate system, as shown in **Figure 12**. The template point cloud is generated by rotating along the x-axis and y-axis of the model point cloud coordinate system with a fixed incremental rotation angle. As shown in **Figure 13a–f**, the template point clouds recorded by the virtual camera are rotated along the y-axis by 0, 60, 120, 180, 240, and 270 degrees, respectively.

After the virtual camera records the template point clouds with different viewing angles, the surface area feature descriptors can be calculated for each template point cloud. First, the OBB of the template point cloud is calculated; it is the smallest 3D rectangular bounding frame that can be covered by the point cloud, as shown in **Figure 14**. The OBB thus obtained can be divided into $k_1 \times k_2 \times k_3$ sub-oriented bounding boxes (Sub-OBB) along its three main directions, as shown in **Figure 15**. The total number of sub-OBB equal to $v = k_1 \times k_2 \times k_3$, and the total number of points covered by the objects in a single sub-OBB is $n_v^p$. If n is the total number of points in the whole template point cloud, then the number of points in a sub-OBB $f_v$ can be expressed as:

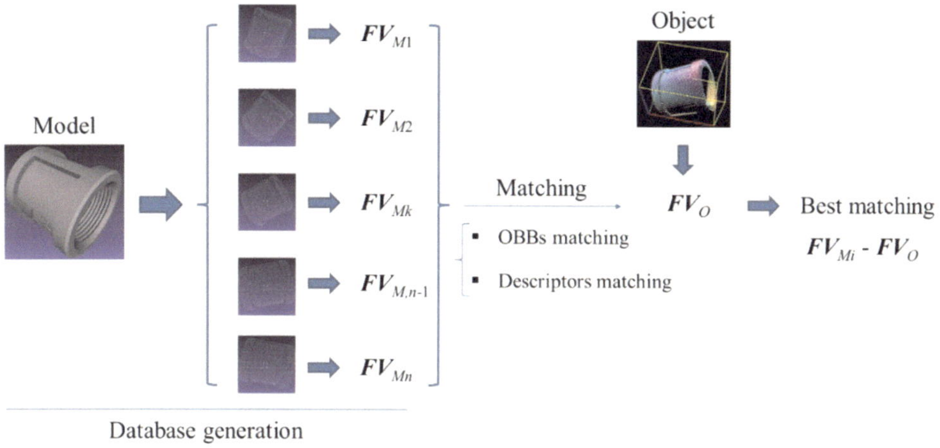


**Figure 11.**
*Matching using regional surface area descriptor (RSAD).*

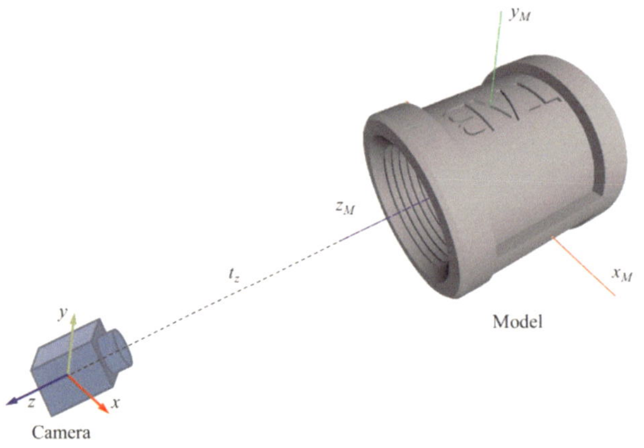


**Figure 12.**
*Location of the virtual camera in model point cloud coordinate system.*

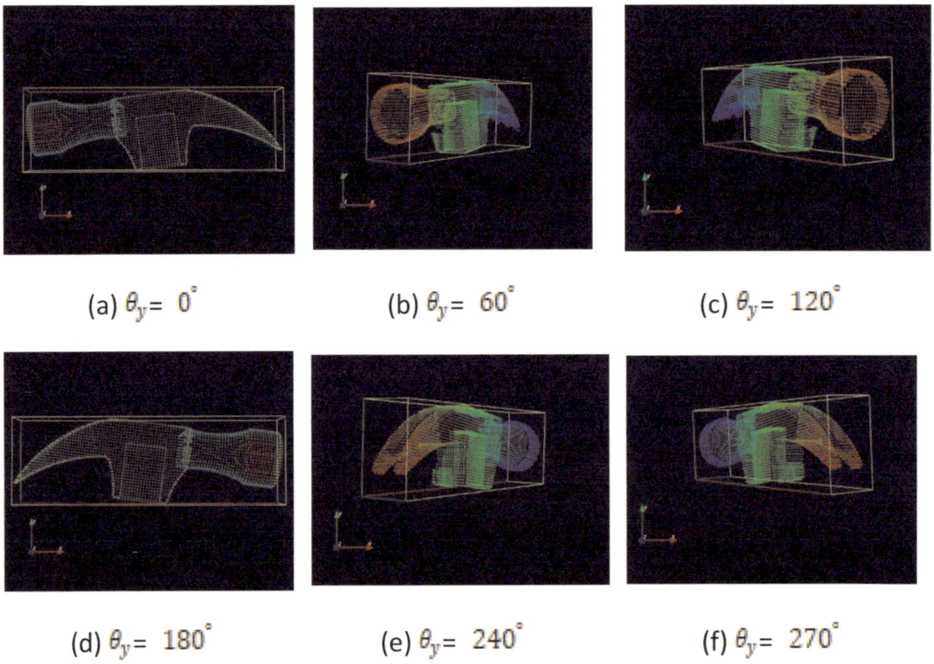


**Figure 13.**
*(a)–(h) are model point clouds captured by the virtual camera at different viewing angles.*

$$f_v = \frac{n_v^p}{n} \tag{2}$$

where $n_v^p$ is the total number of points in the $v_{th}$ sub-OBB and n is the total number of points in the template point cloud.

According to (2), the percentage of point clouds in each sub-OBB can be calculated to obtain the regional surface area distribution of the whole point cloud, which is called the surface area feature descriptor FV and expressed as:

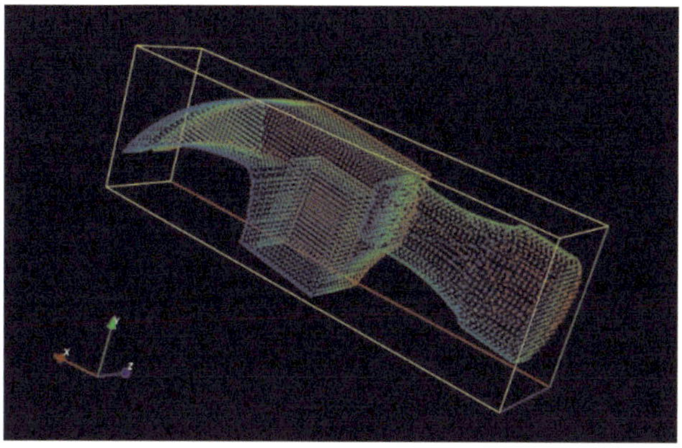


**Figure 14.**
*OBB of an object point cloud.*

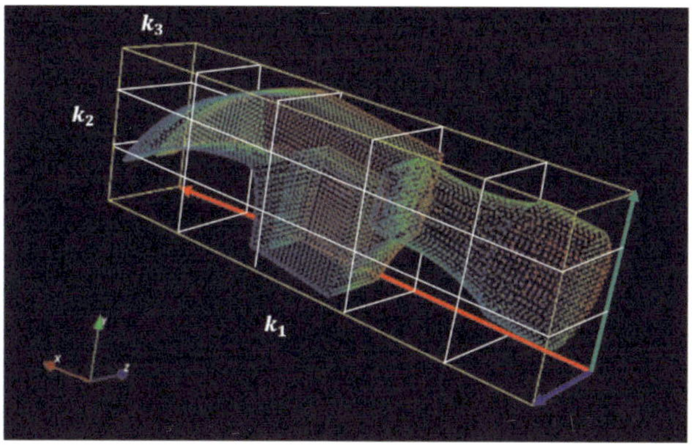


**Figure 15.**
*Sub-OBB of an object point cloud.*

$$FV = \{f_0^p, f_1^p, \dots, f_v^p\} \tag{3}$$

The surface area distribution of the object point cloud in **Figure 14** is shown in **Figure 16**.

### *2.2.2 Feature matching*

Feature matching involves two steps, which aim to obtain the closest pose between the model point cloud and the object point cloud. First, the OBB parameters between the object point cloud and the model point cloud are compared. If the above matching conditions are met, the surface area feature descriptors of the object point cloud and the model point cloud are compared. The feature matching process is summarized in **Figure 17**.

Each OBB is represented by three principal vectors and a corner. The matching process involves finding a template point cloud with an OBB size similar to the object point cloud. These two OBBs should satisfy the following equation:

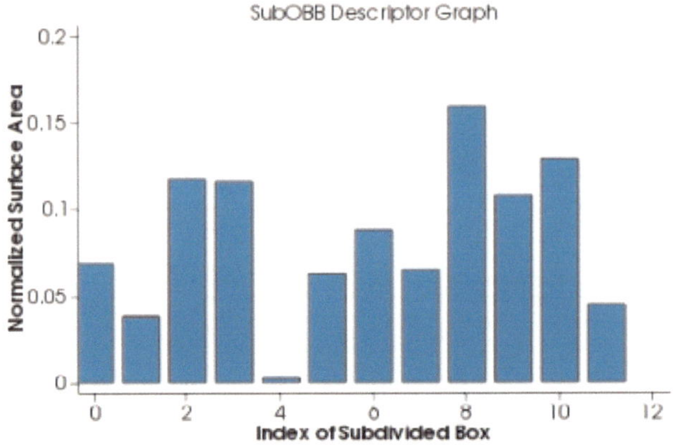


**Figure 16.**
*Regional surface area distribution histogram of the object point cloud.*

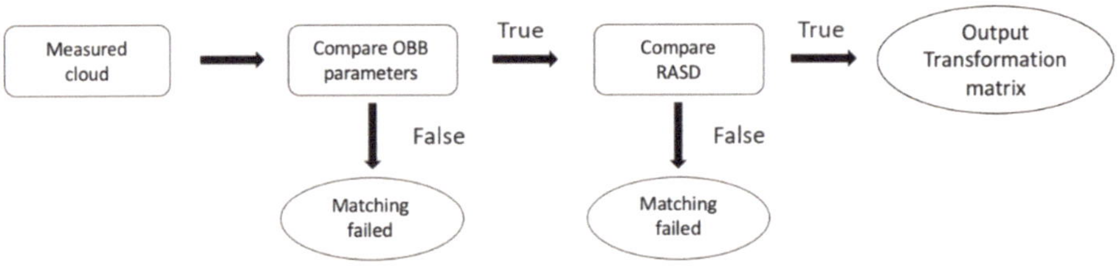


**Figure 17.**
*Flowchart of feature matching process.*

$$d_{corr} = \frac{1}{3}\left\{\frac{\|CC_{M1} - CC_{O1}\|}{\|CC_{M1}\|} + \frac{\|CC_{M2} - CC_{O2}\|}{\|CC_{M2}\|} + \frac{\|CC_{M3} - CC_{O3}\|}{\|CC_{M3}\|}\right\} < d_{thresh} \quad (4)$$

where $d_{thresh}$ is the given adequate threshold, $CC_{Mi}$ and $CC_{Oi}$ (i = 1,2,3) are the three principal vectors of the model OBB and object OBB, respectively.

If (Eq. (4)) is satisfied, matching is successful. On the contrary, if $d_{corr} > d_{thresh}$, matching fails, indicating a significant OBB size difference between the object point cloud and the model point cloud. Assume that **Figure 18a** and **b** depict, respectively, the OBB of the object and model point cloud. According to details of their three

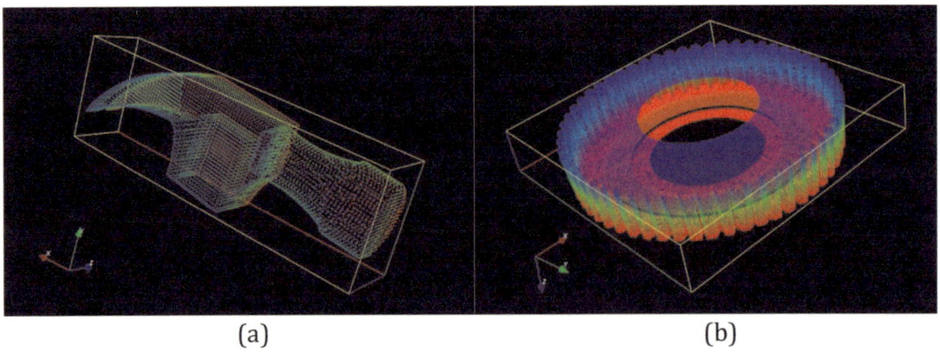


**Figure 18.**
*(a) OBB of the model point cloud and (b) OBB of the object point cloud.*

| | Model point cloud ($CC_M$) | Object point cloud ($CC_O$) |
|---|---|---|
| Principal vector of X-axis | (112.4,0,0) | (69.3,0,0) |
| Principal vector of Y-axis | (0,35.1,0) | (0,68.6,0) |
| Principal vector of Z-axis | (0,0,21.7) | (0,0,12.1) |

**Table 1.**
*OBB parameters of model and object point clouds.*

principal vectors listed in **Table 1**, the OBB matching coefficient $d_{corr}$ between the two point clouds is 0.593. Hence, if the matching coefficient threshold is set as 0.2, then the matching between **Figure 18a** and **b** fails because $d_{corr}$ exceeds the given threshold.

Following this, to determine the best match between the model point cloud and the object point cloud measured using the 3D scanner, the feature descriptors of the model point cloud and the object point clouds measured at different viewing angles are matched using conventional normalized cross-correlation (NCC) [14, 15].

If the OBB matching satisfies (Eq. (4)), then the regional surface area descriptors of the two-point clouds are compared. The regional surface area descriptor shows the surface area distribution of the point cloud within its OBB. The similarity of regional surface area descriptors between the two point clouds can be determined using the NCC method [14, 15]. Assume that the regional surface area descriptor of the object point cloud is expressed as $F_O = \{f_v, v = 0, \ldots, n_v\}$, and the regional surface area descriptor of the model point cloud is $F_M = \{f_v, v = 0, \ldots, n_v\}$, then the NCC coefficient between the two descriptors $F_O$ and $F_M$ can be expressed as:

$$C(F_O, F_M) = \frac{\sum_{v=0}^{n_v}\left(f_v - \bar{f}\right)\left(f_{Mv} - \overline{f_M}\right)}{\sqrt{\sum_{v=0}^{n_v}\left(f_v - \bar{f}\right)^2 \cdot \sum_{v=0}^{n_v}\left(f_{Mv} - \overline{f_M}\right)^2}} \tag{5}$$

where

$$\bar{f} = \frac{1}{n_v + 1}\sum_{v=0}^{n_v} f_v \tag{6}$$

$$\overline{f_M} = \frac{1}{n_v + 1}\sum_{v=0}^{n_v} f_{Mv} \tag{7}$$

If the NCC $C(F_O, F_M)$ between the RASD of the model point cloud and that of the object point cloud is greater than the preset threshold, the two-point clouds have similar poses in the 3D space. The relationship between the two point clouds can be calculated using OBB parameters. Assume that the initial conversion matrix $T_{initial}$ is the relationship between the two point clouds, as shown in **Figure 19**. Their conversion relationship in the space $T_{initial}$ can be expressed as:

$$\begin{bmatrix} x_c & x_c + v_{11} & x_c + v_{21} & x_c + v_{31} \\ y_c & y_c + v_{12} & y_c + v_{22} & y_c + v_{32} \\ z_c & z_c + v_{13} & z_c + v_{23} & z_c + v_{33} \\ 1 & 1 & 1 & 1 \end{bmatrix} = T_{initial} \begin{bmatrix} x_{Mc} & x_{Mc} + v_{M11} & x_{Mc} + v_{M21} & x_{Mc} + v_{M31} \\ y_{Mc} & y_{Mc} + v_{M12} & y_{Mc} + v_{M22} & y_{Mc} + v_{M32} \\ z_{Mc} & z_{Mc} + v_{M13} & z_{Mc} + v_{M23} & z_{Mc} + v_{M33} \\ 1 & 1 & 1 & 1 \end{bmatrix} \tag{8}$$

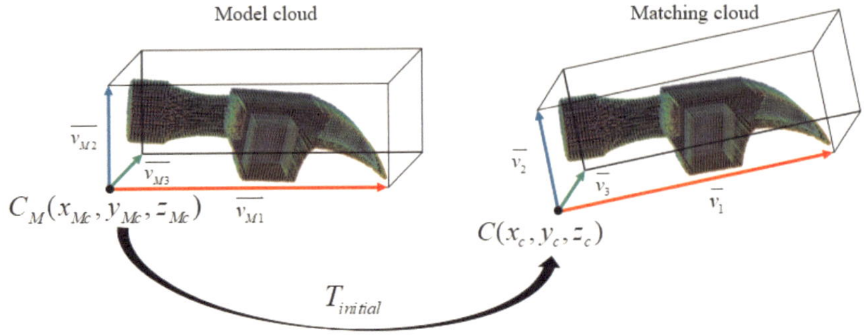


**Figure 19.**
*Schematic diagram of the relationship between object and model point cloud.*

where

$$T_{initial} = \begin{bmatrix} r_{11} & r_{21} & r_{31} & t_x \\ r_{12} & r_{22} & r_{23} & t_y \\ r_{13} & r_{23} & r_{33} & t_z \\ 0 & 0 & 0 & 1 \end{bmatrix}.$$

## 2.3 Fine alignment

The iterative closest point (ICP) algorithm proposed by Besl [7] is a reliable and widely used point-cloud fine alignment algorithm. The ICP algorithm starts with two point clouds and an initial guess for their relative rigid-body transform and refines iteratively the transform by repeatedly generating pairs of corresponding points on the point clouds and minimizing an error metric. The flowchart of the ICP algorithm is shown in **Figure 20**.

To further improve the computational efficiency of the ICP algorithm and optimize the alignment results of the 3D point cloud, many variants have been developed from the basic ICP concept, such as NICP [16], GICP [17], and Point-to-Plane ICP [18]. Rusinkiewicz [19] classified these variants as six stages in the ICP algorithm, illustrated in **Figure 21,** and indicated that the most critical stages that affect the convergence speed and the accuracy of the ICP algorithm are correspondence calculation and error minimization.

In this work, the classic ICP algorithm is further improved with the normal shooting method used in the stage of correspondence calculation instead of finding the closest point. Moreover, in the stage of metric error selection, the point-to-plane distance is applied instead of the point-to-point distance. The concept of the normal shooting method and point-to-plane distance is shown in **Figures 22** and **23**, respectively.

In the proposed approach, test data make up the measured point cloud obtained by measuring a robotic arms end effector with the developed structured-light 3D scanner. In contrast, target data make up the measured point cloud transformed by a known transformation matrix, as shown in **Figure 24a**. The white point cloud is the original point cloud, and the green point cloud contains target data transformed by a known transformation matrix. An example of the end effector alignment result is shown in **Figure 24b**. It is important to indicate that no artifact calibration is deployed

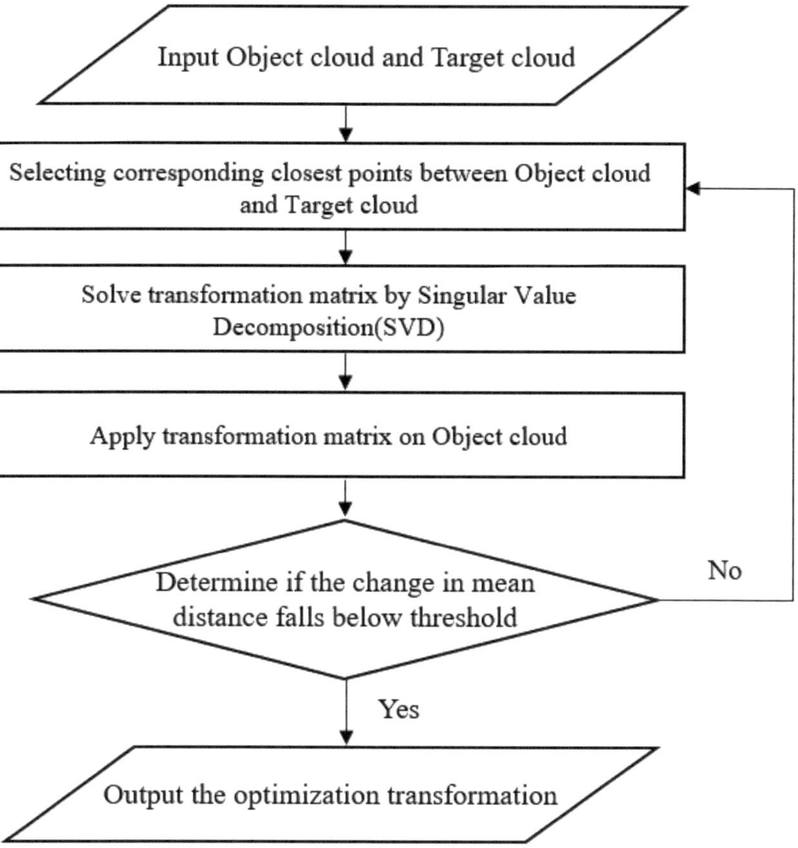


**Figure 20.**
*Flow chart of the ICP algorithm.*

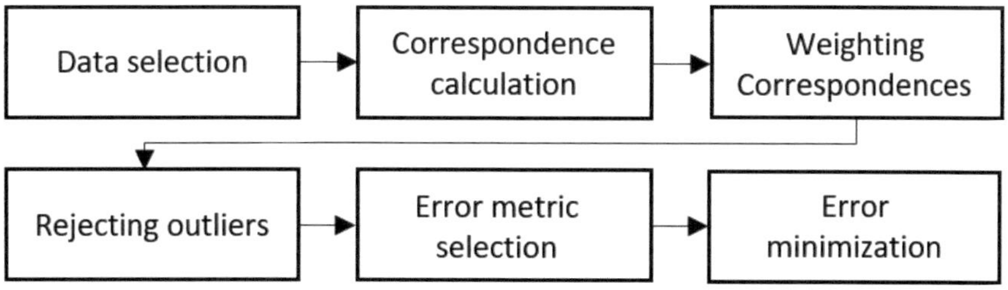


**Figure 21.**
*Six stages of the ICP algorithm.*

in the detection and alignment of the method. This is significantly different from the conventional method, which generally utilizes a known or pre-defined artifact for alignment and calibration purposes. The alignment object used in the method is only the robot arm itself.

**Figure 25** compares the performance between the classic ICP algorithm and the proposed method. As can be seen, the proposed method takes only 0.116 seconds for the mean squared error (MSE) of point-to-point distance to converge to less than

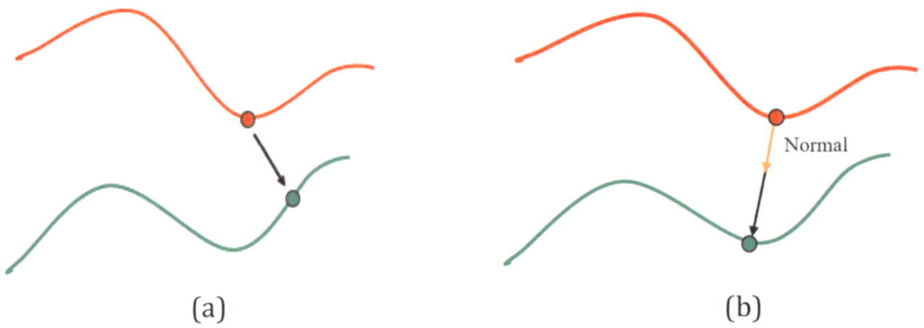


**Figure 22.**
*Correspondence calculation is achieved by (a) the conventional method by finding the closest point and (b) the proposed method using normal shooting.*

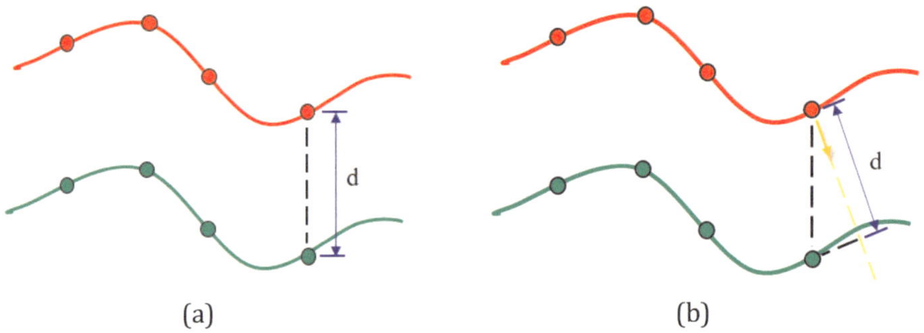


**Figure 23.**
*Error metric selection is achieved by (a) the conventional method using point-to-point distance and (b) the proposed method using point-to-plane distance.*

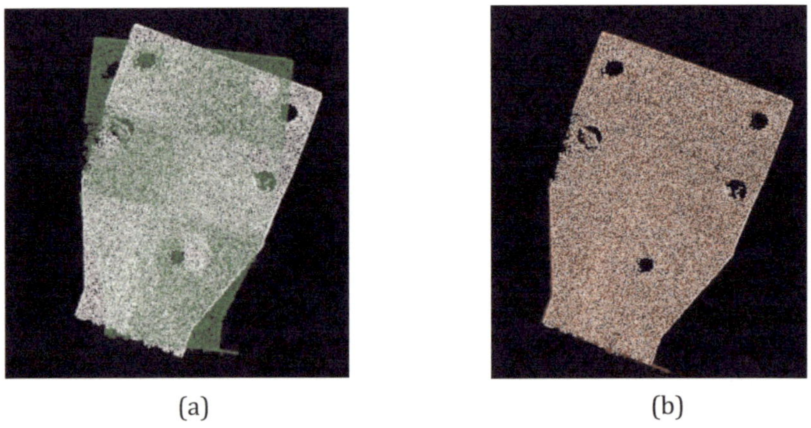


**Figure 24.**
*Fine point-cloud registration: (a) test data (white) and target data (green); (b) registration result after fine alignment.*

1 μm, while the classic ICP takes a longer time of 0.659 seconds to reach convergence, yet a higher deviation of 230 μm. These results indicate that the proposed method has a faster convergence speed and higher accuracy than the classic ICP algorithm.

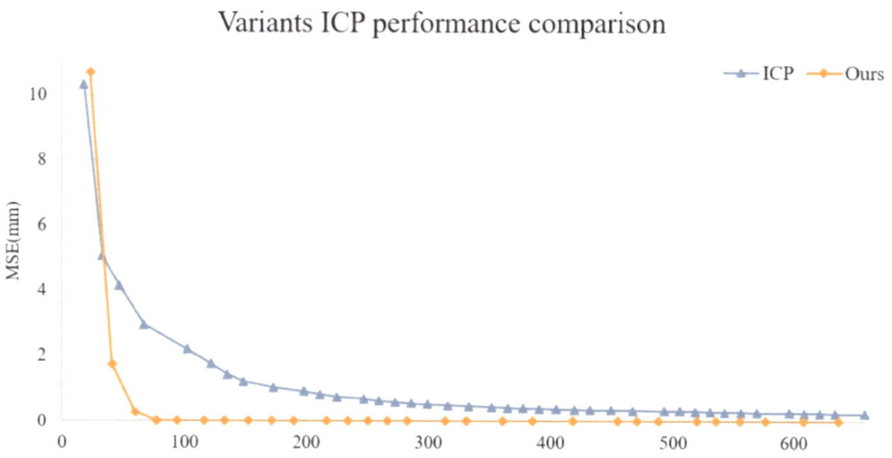


**Figure 25.**
*Comparison of performance between conventional ICP and the proposed method.*

## 3. 6DOF measuring system

### 3.1 Experimental setup

**Figure 26** displays the setup of the developed system for detecting 6DOF variations of the robotic arms end effector, which comprises the developed structured-light 3D scanner, a wafer handling robot, and a computer. The specifications of these three modules are listed in **Table 2**.

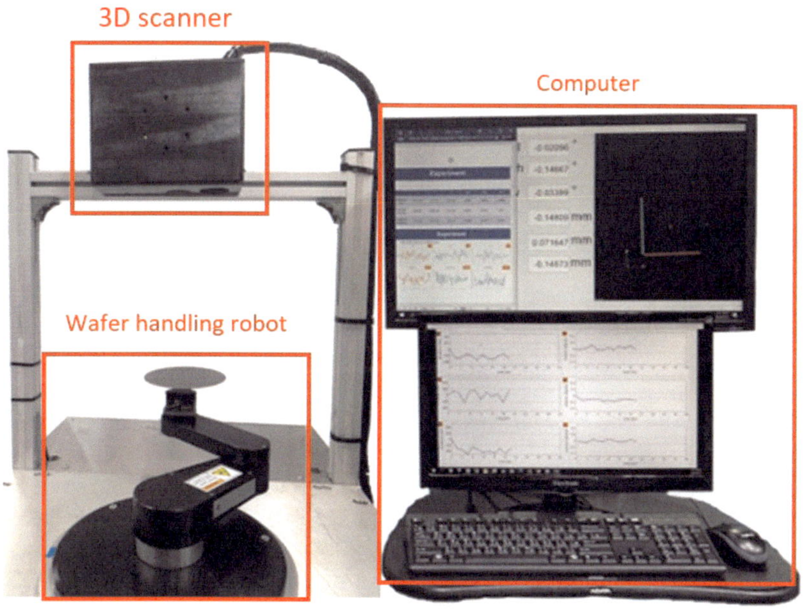


**Figure 26.**
*Experimental setup.*

| | |
|---|---|
| **Wafer handling robot** | |
| Positioning accuracy | ± 25 μm |
| Wafer sizes | 75 to 300 mm |
| Structured light 3D scanner | |
| Working distance | 250 mm |
| Measurable depth range | 100 mm |
| Volumetric measuring accuracy | 60 μm |
| Computer | |
| CPU | Intel Core i7–8700 |
| RAM | 32.0 GB |

**Table 2.**
*System specifications.*

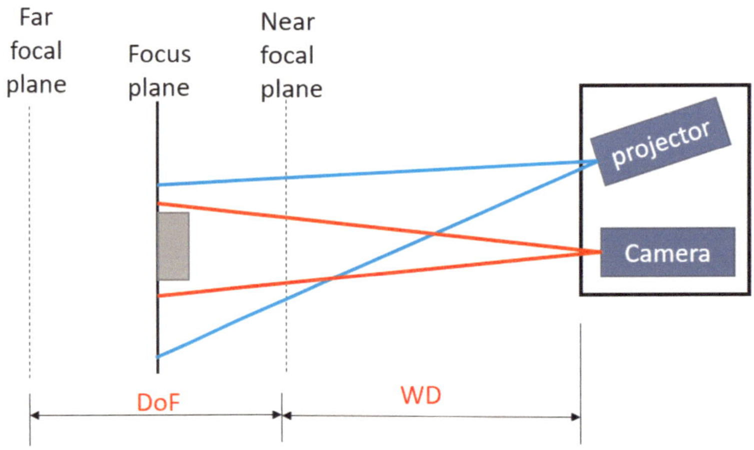


**Figure 27.**
*Schematic diagram of the 3D structured-light measurement probe.*

## 3.2 Design of 3D structured-light measurement probe

The 3D structured-light measurement probe, as shown in **Figure 27**, is designed according to the triangulation measuring principle. As can be seen, it comprises a structured-light projector and a camera, with the projector's optical axis and the camera's optical axis intersecting at a point and forming a triangular relationship with each other. Moreover, the greater the angle between the projector and the camera, the better the depth resolution is. Under a trade-off between the effect of light shielding and depth resolution, this system adopts an included angle of 30°. Moreover, to avoid blurring or poor contrast of the image taken by the sensor module, the measurement depth range is defined as the area where the contrast between the nearest focus surface and the farthest focus surface of the camera lens in the modulation transfer function (MTF) diagram, was maintained above 70%. As shown in **Figure 28**, the area is a trapezoid space, and the camera's field of view is gradually enlarged from the front to the back focal plane.

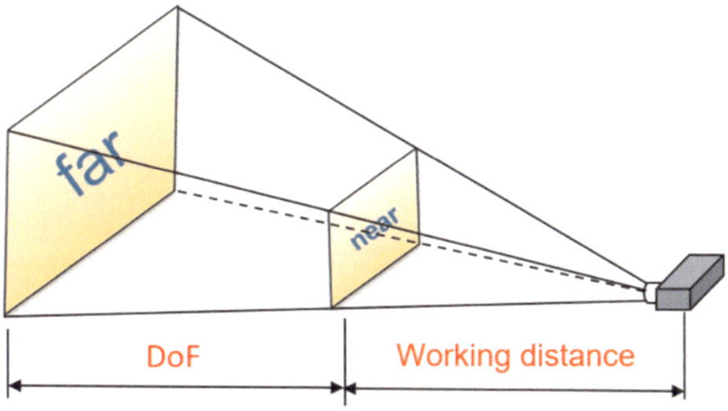


**Figure 28.**
*Schematic diagram of the structured-light probe measurement range.*

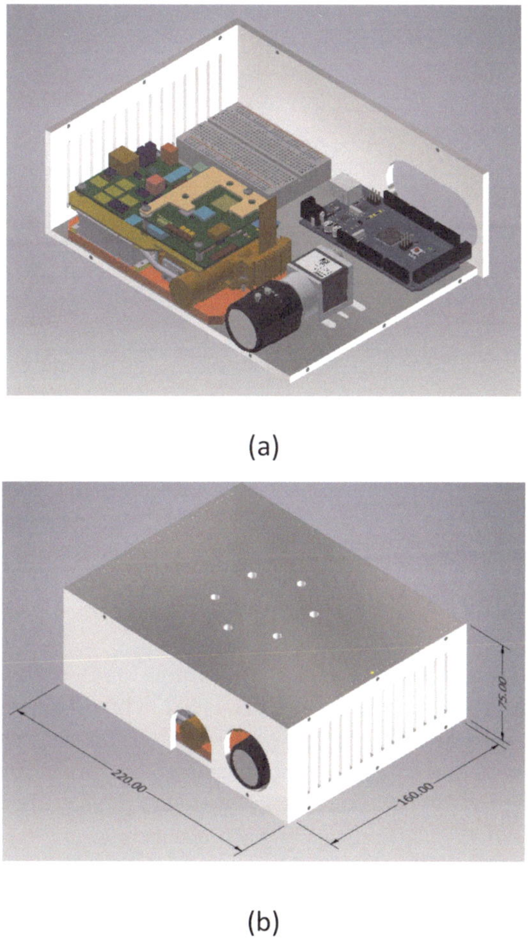


**Figure 29.**
*Engineering drawing of the structured-light probe: (a) internal structure and (b) external view.*

**Figure 29** is the engineering drawing of the structured-light measurement probe designed by Autodesk Inventor. As shown in the actual image (**Figure 30**), the internal structure includes a DLP projection module, an image sensor, a perspective

**Figure 30.**
*Actual internal image of the probe.*

**Figure 31.**
*Actual appearance image of the probe.*

lens, Arduino Mega 2560, and a breadboard. Arduino Mega 2560 is to synchronize the trigger signal between the camera and the projector. The measurement probe's outer shell (**Figure 31**) is made of high-strength aluminum alloy to avoid deformation by external forces. The two side covers of the measurement probe are designed for heat dissipation so that the heat generated by the hardware device can be removed through the heat dissipation holes.

## 4. Experimental results analysis and discussion

Usually, when the robotic arm moves to a specified position, it generally vibrates and shakes due to undesired external force or load. Therefore, upon reaching the working position, it often stops for a short time till the shaking ends and the arm becomes stable. In this study, the end effector of the robotic arm is measured in three dimensions during the standstill time, and the morphological information of the end effector is recorded and compared with the prebuilt database to find the point cloud closest to the pose and obtain the initial transformation matrix. The position and pose variation of the end effector are calculated and displayed as error diagrams in the measurement software's graphic user interface (GUI), as shown in **Figure 32**.

### 4.1 Database generation

Coarse registration requires database creation for prematched objects before measurement. In this experiment, the 3D point cloud information of the robot arm end effector at the working position was used as the model point cloud, the virtual camera recorded the template point clouds at different viewing angles, and the surface area feature descriptors were calculated and stored in the database as the basis for subsequent comparison.

Step 1: Move the robot arm to the working position, as shown in **Figure 33**.

Step 2: Obtain the 3D constructed point cloud of the robotic arms ends effector by the scanning probe. **Figure 34** shows the measurement range of the scanning probe. **Figure 35** shows the image of the robotic arm captured by the camera, and **Figure 36** shows the raw data of the 3D reconstructed point cloud.

Step 3: Preprocess point cloud data. **Figure 37** shows the point cloud after noise removal, and **Figure 38** shows the downsampled point cloud.

Step 4: Create a multiview template point cloud from the model point cloud, as shown in **Figure 39a–h**.

5: Calculate the regional surface area descriptor of the corresponding template point cloud, as shown in **Figure 40a–h**.

### 4.2 *Experimental results analyses and discussion*

To verify the actual performance of the developed method, a robotic arm with an end effector was repeatedly moved 100 times to observe its variations in position and orientation. As shown in **Figure 41a** and **b**, the robotic arm repeatedly moves between position A and position B. In the test, the developed 3D scanner was integrated with

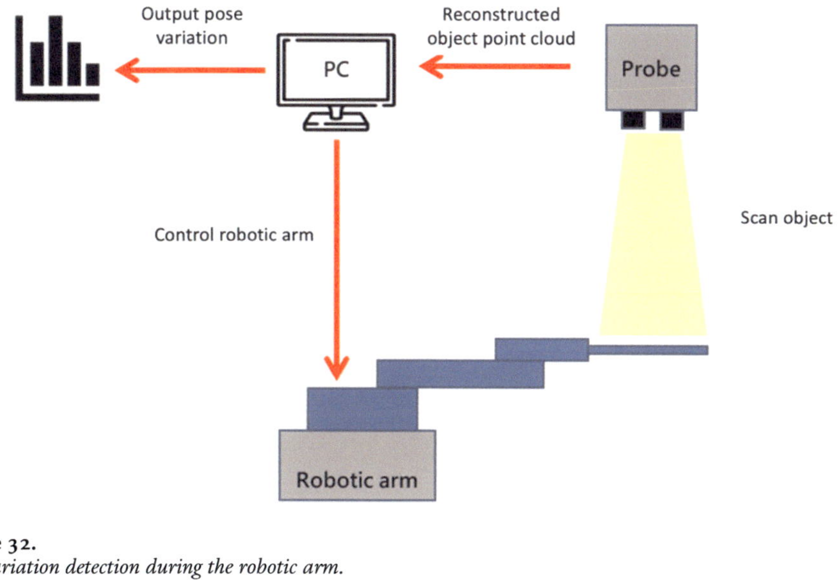


**Figure 32.**
*Pose variation detection during the robotic arm.*

**Figure 33.**
*The robot arm is moved to the work position.*

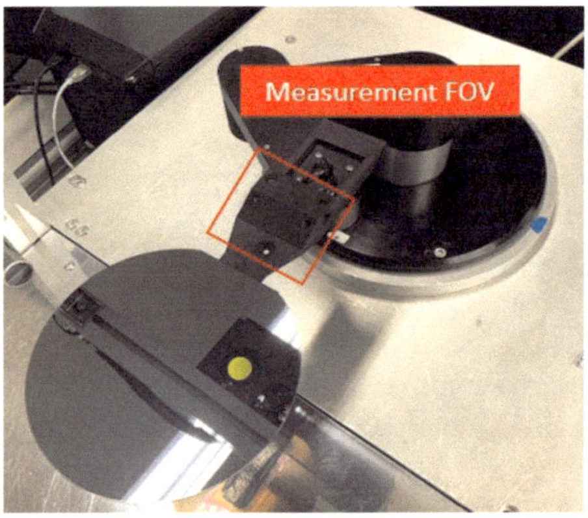


**Figure 34.**
*Measurement FOV of the probe.*

**Figure 35.**
*Image of robot arm end effector captured by the measurement probe.*

**Figure 36.**
*Reconstructed point cloud of robot arm end effector.*

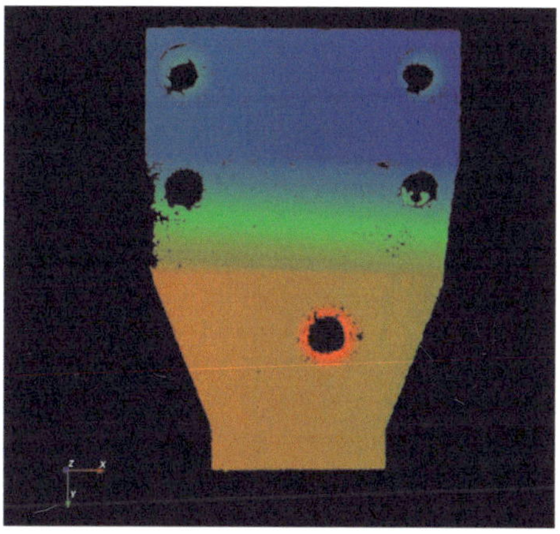

**Figure 37.**
*Point cloud after outlier removal.*

the system to measure the robotic arms end effector for its 3D point clouds whenever the robotic arm reaches position B.

In the test, the 6DOF variations of the robotic arms end effector are compared with the measured point cloud whenever the arm moves from position A to B and stops at position B. **Figure 42a** shows the image captured using the developed 3D scanner when the robotic arm reaches position B and **Figure 42b** is the original reconstructed point cloud. **Figure 43** shows point-cloud registration results before and after alignment without referencing any target.

**Figures 44** and **45** show the dynamic variations in position and angular orientation, respectively, obtained after 100 tests. As can be seen, for the robot end effector in the x-, y-, and z-axis, the averaged pose variations are 0.039 mm, 0.003 mm, and 0.005 mm, respectively. In contrast, the averaged orientation variations are 0.009°, 0.029°, and 0.009°, respectively. These results indicate that the tested robot end effector achieved positioning with micron accuracy in the worst scenario. The

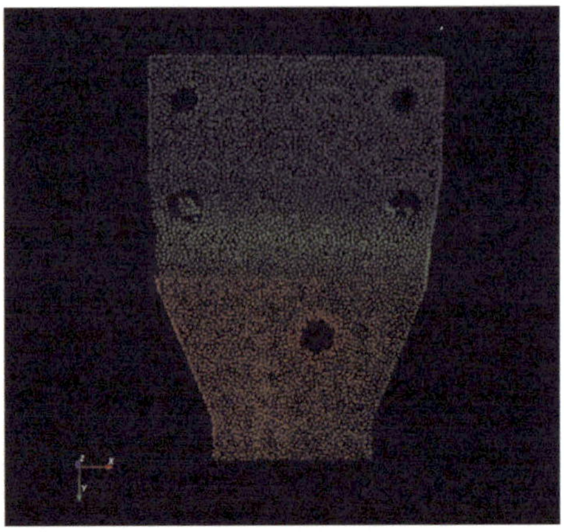

**Figure 38.**
*Point cloud after downsampling.*

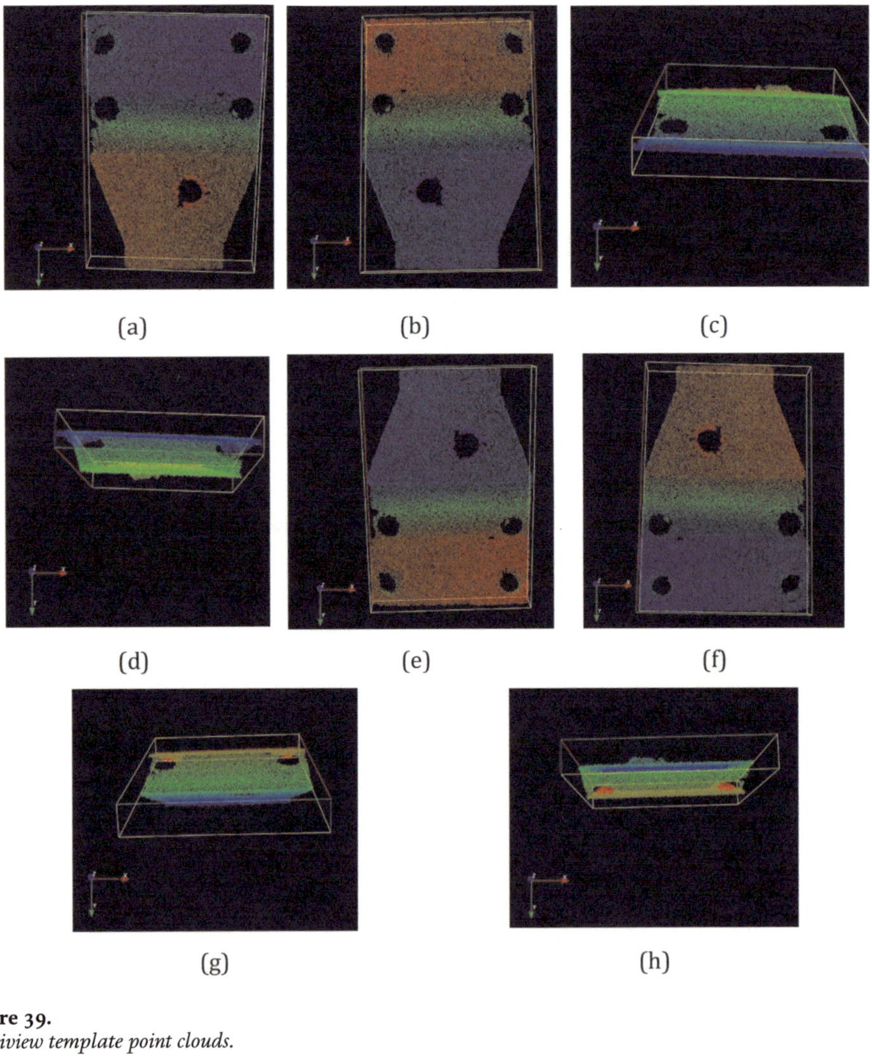

(a) (b) (c) (d) (e) (f) (g) (h)

**Figure 39.**
*Multiview template point clouds.*

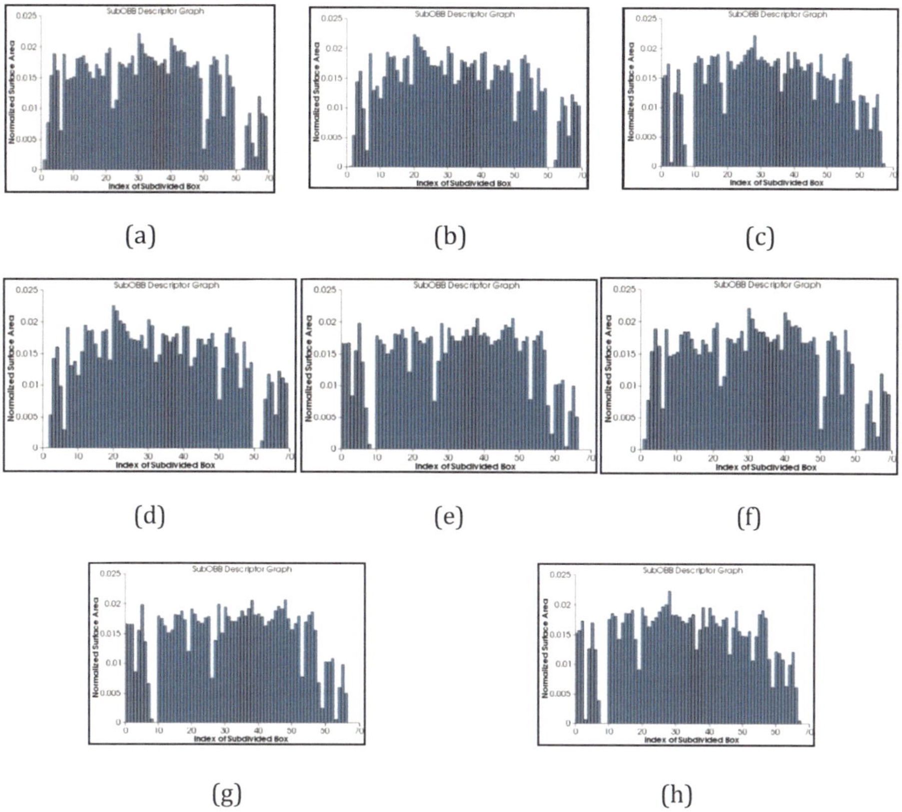


**Figure 40.**
*RSAD of the corresponding template point cloud.*

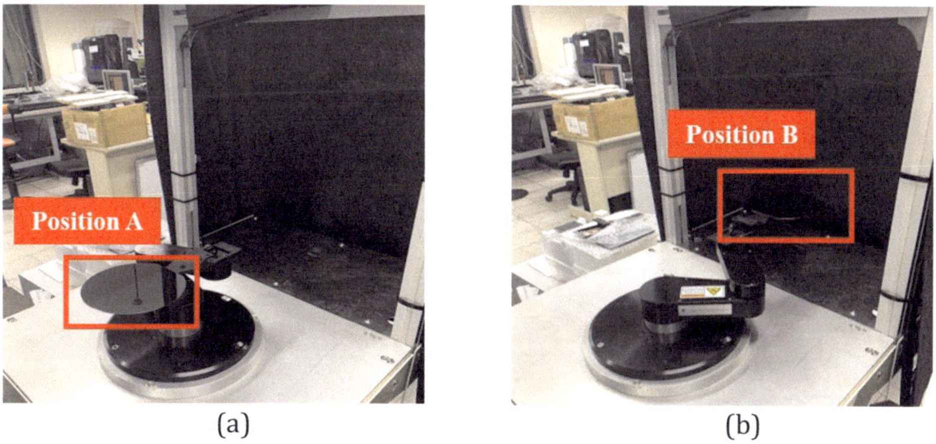


**Figure 41.**
*Different measured positions of the tested robotic arm with its end effector.*

experimental results also show that the translational error of the x-axis and the rotation error of the y-axis may increase significantly with the operation time of the robotic arm. This discovery can lead to possible inspection and maintenance of the

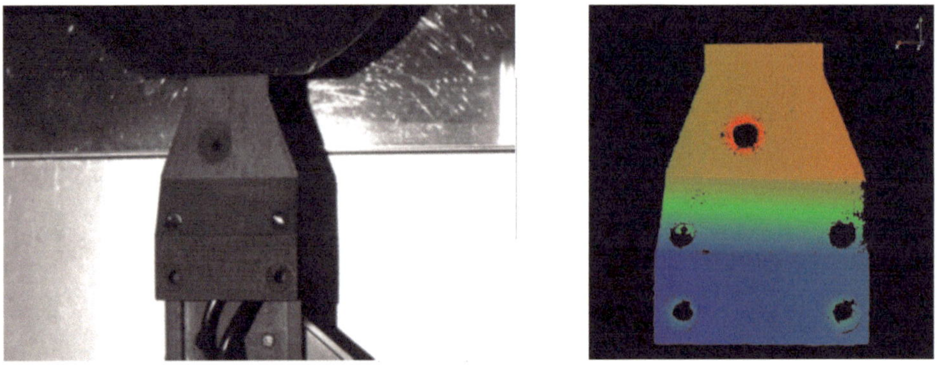


**Figure 42.**
*(a) Image captured using the developed 3D scanner when the robotic arm reaches position B. (b) Reconstructed point cloud at position B.*

**Figure 43.**
*Point-cloud registration before (left) and after (right) alignment without using any calibration artifact.*

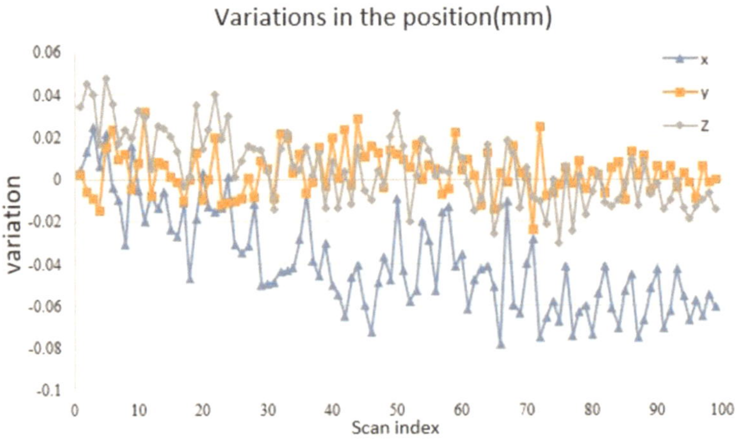


**Figure 44.**
*Dynamic variations in positioning of robot end-effector along x-, y- and z axes.*

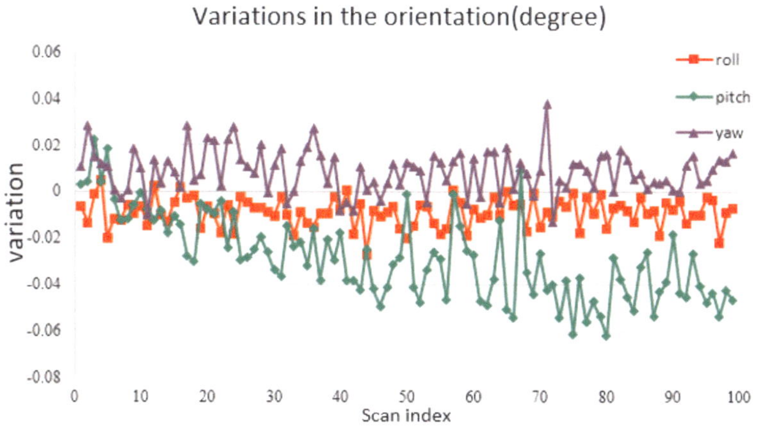


**Figure 45.**
*Dynamic variations in angular orientation of the robot end effector along x-, y- and z axes, defined as roll, pitch, and yaw angular errors.*

robot to be arranged to ensure robust manufacturing operation and avoid any potential catastrophic damage.

### 4.3 Error analysis of robot arm pose variation

The sources of error in the analysis of the robot's pose variation include the error of the structured-light measurement probe, pose detection algorithm, and the robotic arm itself. In addition to the positioning error of the robot arm itself, which is known from the technical specifications provided by the original manufacturer, the positioning repetition of the robot arm is 25 μm. In contrast, the reconstruction error of the measurement probe and the error of the pose detection algorithm must be found through experiments. The ideal point cloud data is used as the benchmark to quantify how error sources affect measurement results. The principle is to transform the known point cloud data through a known transformation matrix and to obtain the pose variation using the proposed pose detection algorithm. The method obtains the transformation matrix between the transformed point cloud data and the original point cloud data. It compares the known transformation matrix with the transformation matrix obtained using the proposed algorithm to quantify the error. The error of the structured-light measurement probe is calculated by repeatedly measuring the pose variation of the robot arm at a fixed position, and the remaining error after deduction from the algorithm error can be regarded as the error of the structured-light measurement probe.

The translation error $T_{err}$, defined as the absolute difference between the translation component of the known transformation matrix $T_{true}$ and the translation vector of the transformation matrix $T_{alg}$ obtained using the proposed pose detection algorithm [20], is expressed as:

$$T_{err} = \left\| T_{alg} - T_{true} \right\| \tag{9}$$

The measured point cloud is converted in the 3D space to quantify the algorithm error using a known transformation matrix, as shown in **Figure 46**. The object to be

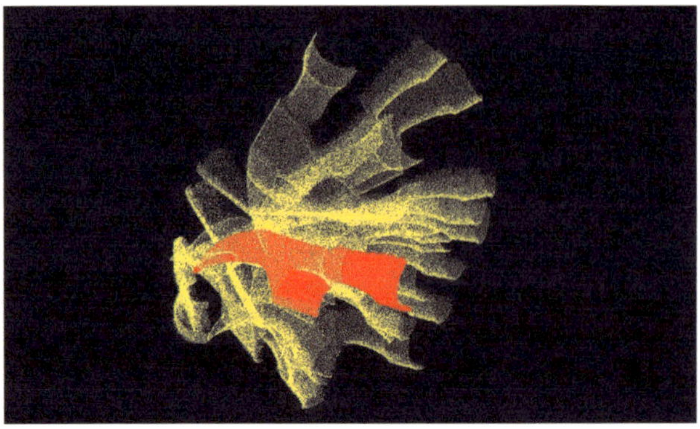

**Figure 46.**
*Original (red) and transformed (yellow) point clouds.*

measured is a 3D printing hammer. The red point cloud is the original point cloud, while the yellow one is transformed from the original point cloud in 3D space using 30 different known transformation matrices. **Figure 47** shows the translation errors obtained using the proposed pose detection algorithm. The average translation error of the 30 transformations is 0.007 mm.

System uncertainty of the 3D structured-light measurement probe will cause the measurement results of the same object to deviate. The robotic arm was scanned by the measurement probe 30 times to detect pose variation. For a fixed robot arm, the pose variation should be zero. Otherwise, the measurement obtained after deducting the algorithm error can be regarded as an error attributed to the structured-light measurement probe. **Figure 48** shows the translation error obtained for the 30 scannings. The average translation error contributed by the structured-light measurement probe was 46 μm.

**Table 3** summarizes the error source distribution of the robot arm pose variations. As can be seen, the error comes mainly from the structured-light measurement probe. Light perception by the image sensor in the measurement probe may vary for the same object even under the same conditions, thus causing differences in measurement results. Experiments detected an average pose variation of 46 μm from this source, accounting for 58.5% of the total error. The next major source of error is the robotic

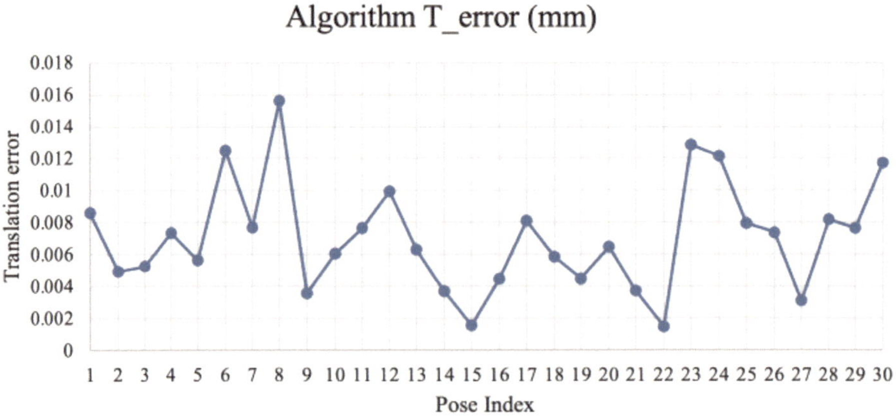


**Figure 47.**
*Translation error was obtained using the proposed pose detection algorithm.*

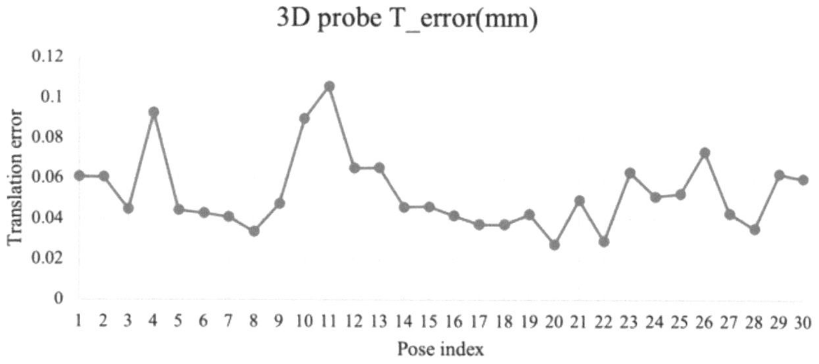


**Figure 48.**
*Translation error contributed by the structured-light measurement probe.*

| | Error (μm) | Percentage (%) |
|---|---|---|
| Measurement probe | 46 | 58.5 |
| Robotic arm | 25 | 31.8 |
| Pose detection algorithm | 7.6 | 9.6 |
| Total | 78.6 | 100 |

**Table 3.**
*Error budget analysis of pose variation of the robotic arm.*

arm itself. System errors lead to the difference in the position specified by the controller and the actual position reached. According to the original technical specifications, the pose variation is 25 μm, which accounts for 31.8% of the total error. A minor source of error is the pose detection algorithm, with pose variation attributed to floating point and matrix calculation errors. An average pose variation of 7 μm was detected, accounting for 9.6% of the total error.

## 5. Conclusions

This chapter introduces a novel method for measuring 6DOF variations of the robotic arms end effector using regional surface area descriptors and variant ICP to monitor accurate positioning and orientation. Experimental test results confirmed that the proposed method can accurately position and detect three angular errors, namely pitch, yaw, and roll angle, without referencing any known artifact for system alignment. A comprehensive measurement uncertainty was also performed to identify possible measured sources. From the measurement results, it reveals that the 3D optical probe can achieve 60 μm measurement accuracy and 100 μm depth resolution within the entire measurement range of 100 mm. On the other hand, for the automatic detection of the position and orientation on the robot arm end-effector, the experimental results show that it achieves the positioning accuracy of 100 um in the X, Y, and Z positions within the entire measurement range. Most importantly, it was also verified that the angular detection accuracy can be kept within 0.01 degrees for the pitch, yaw, and roll angular motions. The developed method can be widely applied to precise 6DOF detection of arbitrary objects moving dynamically in 3D space. Its high-accuracy 6DOF monitoring with target-free 3D image registration capability is a significant technical breakthrough in automated optical inspection (AOI) and precision metrology in manufacturing.

## Author details

Liang-Chia Chen*, Sheng-Hao Huang and Bo-Han Huang
Department of Mechanical Engineering, National Taiwan University, Taipei, Taiwan

*Address all correspondence to: lchen@ntu.edu.tw

## References

[1] Cong M, Zhou Y, Jiang Y, Kang R, Guo D. An automated wafer-handling system based on the integrated circuit equipments. In: 2005 IEEE International Conference on Robotics and Biomimetics–ROBIO. Hong Kong, New York: IEEE; 2005. pp. 240-245

[2] Huang PW, Chung KJ. The prediction of positioning shift for a Robot arm using machine learning techniques. In: 2019 14th International Microsystems, Packaging, Assembly and Circuits Technology Conference (IMPACT). Taipei, New York: IEEE; 2019. pp. 58-61

[3] Huang PW, Chung KJ. Task failure prediction for wafer-handling robotic arms by using various machine learning algorithms. Measurement and Control. 2021;**54**:701-710

[4] Johnson AE, Hebert M. Using spin images for efficient object recognition in cluttered 3D scenes. IEEE Transactions on Pattern Analysis and Machine Intelligence. 1999;**21**:433-449

[5] Chua CS, Jarvis R. Point signatures: A new representation for 3d object recognition. International Journal of Computer Vision. 1997;**25**:63-85

[6] Hoang DC, Chen LC, Nguyen TH. Sub-OBB based object recognition and localization algorithm using range images. Measurement Science and Technology. 2016;**28**:025041

[7] Besl PJ, McKay ND. Method for registration of 3-D shapes. In: Sensor Fusion IV: Control Paradigms and Data Structures. Boston, Washington: SPIE; 1991. pp. 586-606

[8] Geng J. Structured-light 3D surface imaging: A tutorial. Advances in Optics and Photonics. 2011;**3**:128-160

[9] Rusu RB, Blodow N, Marton Z, Soos A, Beetz M. Towards 3D object maps for autonomous household robots. In: 2007 IEEE/RSJ International Conference on Intelligent Robots and Systems. San Diego, New York: IEEE; 2007. pp. 3191-3198

[10] Surynková P. Surface reconstruction. In: Proceedings of the 18th Annual Conference of Doctoral Students-WDS. Prague: MatfyPress; 2009. pp. 204-209

[11] Graham RL, Yao FF. Finding the convex hull of a simple polygon. Journal of Algorithms. 1983;**4**:324-331

[12] Mencl R, Muller H. Interpolation and approximation of surfaces from three-dimensional scattered data points. In: Scientific Visualization Conference (dagstuhl '97). Dagstuhl, New York: IEEE; 1997. p. 223

[13] Gottschalk S, Lin MC, Manocha D. OBBTree: A hierarchical structure for rapid interference detection. In: Proceedings of the 23rd Annual Conference on Computer Graphics and Interactive Techniques. New Oreland, New York: ACM; 1996. pp. 171-180

[14] Zhao F, Huang Q, Gao W. Fast normalized cross-correlation. In: 2006 IEEE International Conference on Acoustics Speech and Signal Processing Proceedings. Toulouse, New York: IEEE; 2006

[15] Yoo JC, Han TH. Fast normalized cross-correlation. Circuits, Systems and Signal Processing. 2009;**28**:819-843

[16] Serafin J, Grisetti G. NICP: dense normal based point cloud registration. In: 2015 IEEE/RSJ International Conference on Intelligent Robots and

Systems (IROS). Hamburg, New York: IEEE; 2015. pp. 742-749

[17] Segal A, Haehnel D, Thrun S. Generalized-icp. In: Robotics: Science and Systems V. Seattle, Cambridge: MIT Press; 2009. p. 435

[18] Chen Y, Medioni G. Object modelling by registration of multiple range images. Image and Vision Computing. 1992;**10**:145-155

[19] Rusinkiewicz S, Levoy M. Efficient variants of the ICP algorithm. In: Proceedings Third International Conference on 3-D Digital Imaging and Modeling. Quebec City, New York: IEEE; 2001, 2011. pp. 145-152

[20] Choi C, Taguchi Y, Tuzel O, Liu MY, Ramalingam S. Voting-based pose estimation for robotic assembly using a 3D sensor. In: 2012 IEEE International Conference on Robotics and Automation. Saint Paul, New York: IEEE; 2012. pp. 1724-1731

Chapter 7

# A Cognitive Digital-Optical Architecture for Object Recognition Applications in Remote Sensing

*Ioannis Kypraios*


## Abstract

From coastal landscapes to biodiversity remote sensing can on the one hand capture all the natural heritage elements and on the other hand can help in maintaining protected species. In a typical remote sensing application, a few thousands of super high-resolution images are captured and need to be processed. The next step of the processing involves converting those images to an appropriate format for visual display of the data. Then, the image analyst needs to define the regions of interests (ROIs) in each captured image. Next, ROIs need to be defined for identifying specific objects or extracting the required information. First drawback of this processing cycle is the use of image analysis tools which provide them only with scaling or zooming features. Second, there is no conceptual connection between the image analysis tools and the actual processing cycle. Third, such existing tools do not usually automate any steps in the processing cycle. We combine an optical correlator with a supervised or an unsupervised classifier learning algorithm and show how our proposed novel cognitive architecture is conceptually connected with the image analysis processing cycle. We test the architecture with captured images and describe how it can automate the processing cycle.


**Keywords:** object recognition, cognitive digital-optical architecture, image analysis, knowledge representation and learning, remote sensing

## 1. Introduction

Cultural heritage consists of all the tangible and intangible elements from monuments and cultural traditions to natural landscape, including all the extinct or current biological species. Cultural heritage and specifically tourism activities related to cultural heritage contribute to economic growth, regeneration, education and tourism [1, 2]. A previous report [3] published by the HLF and Visit Britain revealed that the heritage tourism is a £12.4 billion a year industry. £7.3 billion of heritage expenditure is based on visits to built heritage attractions and museums, with the overall £12.4 billion including visits to parks and countryside as well. In a recent article, it was reported that to reduce the risk of extinction for all the threatened species worldwide would cost annually approximately £2.97bn, with an additional £47.4bn required per year to establish and manage protected areas for species known to be at risk from habitat loss, hunting and other human activities [4].

Remote sensing has become one of the main technologies nowadays used for protecting cultural heritage due to its non-invasiveness [5]. Thus, remote sensing has been used in many cases for the conservation analysis of monuments, archaeological site detection and risk protection, together with the protection of natural landscapes consisting of living species. Aerial photography is one of the forms of modern remote sensing. It has been applied in many application areas [6]. For example, aerial photography has been successfully used in locating ancient civilisation structures amongst thick jungle vegetation [7].

One of the key tools in remote sensing image analysis is the object recognition. Advanced machine learning (ML) and artificial intelligence (AI) algorithms can be used for detecting and classifying different classes of objects in different applications, such as environmental monitoring, geological hazard detection, land-use/land-cover (LULC) mapping, geographic information systems (GIS), precision agriculture and urban planning. Still, object recognition in remote sensing images (RSIs) can be very challenging due to the large variations in the visual appearance of objects caused by camera viewpoint variations, occlusion, background clutter, and illumination changes. Thus, in low spatial resolution satellite images such as Landsat the recognition of objects becomes even harder. Therefore, higher resolution satellite images such as IKONOS or Quickbird are preferred since provide researchers and image analysts with more detailed spatial and textural information. In effect, a greater range of object categories can be recognised due to the increased sub-meter resolution.

Nowadays the technological advancements in satellite and aerial RSIs have offered us opportunities for new applications in many image analysis areas. Pixels in an image can be grouped and clustered into regions of interest (ROIs) where then object recognition algorithms can be applied for classifying a range of categories of objects. Supervised and unsupervised such classifiers can be utilised depending on the application. Supervised classification require larger amount of data than unsupervised with manual annotation and labelling of a bounding box which contains the true class object for training purposes. However, such manual annotation suffers from scaling up issues when a very large amount of such data is needed. Moreover, it becomes a significantly difficult task either when the ROIs consist of a cluster of a few pixels in an RSI or the objects are occluded, camouflaged and may include complex textures.

Thus, object recognition algorithms become of great importance in applications of RSIs. Extracting the features of an object either through manually labelling them or through a classifier algorithm becomes essential when aiming to recognise them in image analysis applications. However, when dealing with large or very large (big) data, then this task turns complex with high computational costs for processing those extracted feature vectors. The manual labelling is time consuming and unreliable since duplicates or redundant features are often created. Here, we will be focusing more on automated feature extraction through a classifier algorithm for large or very large data.

In Section 2, we discuss the object recognition systems and their design characteristics. In Section 3, we explain the k-means algorithm. Section 4 discusses the optical correlator classifiers. In Section 5, our cognitive object recognition architecture is described. Then, in Section 6 we discuss the results recorded when applying our cognitive architecture. Section 7 contains the Conclusions and future work.

## 2. Object recognition systems for automated image analysis

Aerial or satellite survey consists of several steps that need to be followed. Starting from the data acquisition procedures, they include visual observations, the capturing of imagery and the use of metric measurements on the acquired images. Then, this raw information is typically used to produce a set of documents which consists of text and related images. However, the resulting documents are often hard to use since they have been created for a specific application and require expert knowledge to comprehend. In addition, image quality can vary depending on the remote sensing site conditions. Other challenges on image analysis from aerial surveys can include poor lighting conditions, occlusions, and varying depths. Also, limitations on the geospatial information provided with the acquired images can limit their spatial analysis. Restrictions on the field-of-view (FOV) and the image sensor footprint can affect the number of images collected per aerial survey. Then, consequently, a higher number of images require a higher processing time to be completed. Where time of completion is of essence for cost reasons, then someone can expect that automated image analysis tools are necessary for minimising those processing and completion times.

Integral part of such automated image analysis tools plays the object recognition system used which need to be designed in novel architectures, if they are to be applied for solving more complex problems [8]. Recent advances have led to the development of biologically-inspired also known as cognitive architectures [9] of object recognition systems with separate design blocks of a recognition unit and a separate knowledge learning unit [10, 11]. Thus, cognitive architectures need to exploit the non-linearity, the learning and adaptation to the input data, and provide an attentional mechanism for the hybrid system to be able to select certain input information to be included in its learning against other input. Therefore, knowledge representation and learning becomes a central issue in the design and implementation of such hybrid biologically-inspired pattern recognition [12]. Knowledge representation can have altering effects on problem knowledge learning and problem solving [13]. A problem solving system consists of a domain theory which specifies the task to be solved, the initial problem states and the targeted problem goals, and a control knowledge which guides the decision-making process. Thus, knowledge representation can have a direct effect to the efficiency of the problem solving process [14, 15].

### 2.1 Defining the problem

In this chapter, we describe a new object recognition architecture for improving the speciation of an endangered bird species from aerial surveys. For reasons of confidentiality, we use here the term endangered bird sub-species 1 or, simply, endangered species 1, and the term endangered sub-species 2 or, simply, endangered species 2. In particular, we are looking to improve the accuracy and precision of the recognition of endangered bird sub-species 1and 2 in winter plumage which the task of correctly speciating them becomes harder than in their summer plumage. Our proposed cognitive object recognition architecture incorporates the features of shape, size and colour (3-bands R, G and B) of the endangered sub-species 1 and 2 in the architecture's knowledge representation, and then apply an unsupervised knowledge learning unit for improving the accuracy and precision in recognising them.

## 3. Unsupervised clustering

Unsupervised machine learning refers to learning the input but without a reference to known or labelled data. Unlike to supervised machine learning, unsupervised machine learning algorithms cannot be directly applied to a classification problem since there is no prior knowledge of either the number of object classes or each class threshold. Instead, unsupervised learning can be used for discovering the underlying structure or pattern of the data. Thus, the term "*clustering*" refers to a process of grouping similar things together [14]. Therefore, unsupervised learning can be used for discovering such clusters in the input data.

K-means is an unsupervised algorithm for clustering m objects into k clusters in which each observation belongs to the cluster with the nearest mean [15]. Each centroid is a point in a 2- or N-dimensional space that represents the centre of the cluster. **Figure 1** shows an example of K-means clustering algorithm [17]. The algorithm begins with k randomly placed centroids and assigns every item to the nearest one. After the initial assignment, the centroids start being moved to the average location of all the nodes assigned to them, and new assignments of objects to centroids are redone. The process repeats until the centroids stop being moved by the algorithm [18].

Bennet et al. [19] described a method for unsupervised classification in multitemporal optical RSIs based on discreet wavelet transform (DWT) feature extraction and K-means clustering is proposed. After pre-processing the optical image, they applied a feature extraction using the DWT for creating the input vectors. Then, the authors applied a feature reduction for selecting the most discriminative features using an energy based selection. Finally, they used K-means clustering for unsupervised learning of the input data clusters and compared the results by labelling the clusters using ground truth data. Shulei Wu et al. [20] introduced a novel classification method based on K-means using hue, saturation, value (HSV) colour features. Their novel method with HSV data produced higher classification accuracy results when tested with Landsat satellite data than K-means method with RGB data. Abbas et al. [21] compared K-means unsupervised clustering method with the Iterative Self-Organising Data Analysis Technique Algorithm (ISODATA) unsupervised method for automatically grouping pixels of similar spectral features from remote sensing images.

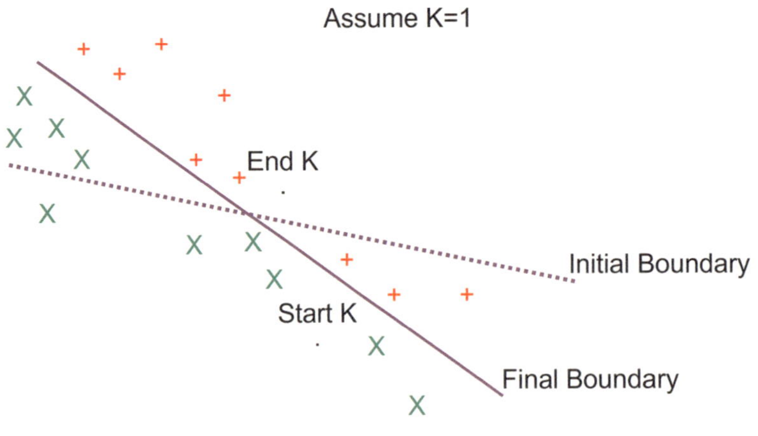


**Figure 1.**
*Example of K-means clustering algorithm. There are initially k = {k1} centroids assigned in the object M-dimensional space. As the algorithm goes through the recursive steps the centroids are re-assigned and the clusters boundaries are moved around the objects space. (adapted by [16]).*

*DOI: http://dx.doi.org/10.5772/intechopen.109028*

Vishwanath et al. [22] combined K-means unsupervised clustering with a Laplacian-of-Gaussian and a Prewitt filters for improving the classification and road edge detection in RSIs. Yin et al. [23] applied K-means clustering algorithm on Lidar based 3D object detection and classification tasks in automated driving (AD). Specifically, they used K-means for 3D points cloud segmentation. The authors reported a high-speed 3D object recognition when run using a GPU enabled platform. Huu Thu Nguyen et al. [24] combined deep learning algorithms with K-means clustering for achieving multiple object detection in both sonar images and 3D point cloud Lidar data.

**Figure 2** shows the K-means algorithm flowchart [25]. The algorithm is a recursive one where previous steps in the flowchart will be called in another step afterwards. The first basic step of this recursive-type algorithm is to determine the number of clusters K. We assume the centroid of these clusters, which it can be any random objects. Alternatively, we can assign as the initial centroids to be the first K objects in sequence. The algorithm as shown on **Figure 2** consists of the following three recursive steps:

1. Determine the centroid coordinate.

2. Determine the Euclidean distance of each object to the centroids.

3. Group the object based on minimum Euclidean distance values.

4. Repeat steps 1–3 till all the centroids stop being moved in the objects space i.e. the algorithm has converged to a solution.

Euclidean Distance d between two points $p_1(x_1, y_1)$ and $p_2(x_2, y_2)$ in X-Y two-dimensional (2D) space is given by:

$$d(p_1, p_2) = \sqrt{(x_2 - x_1)^2 + (y_2 - y_1)^2} \quad (1)$$

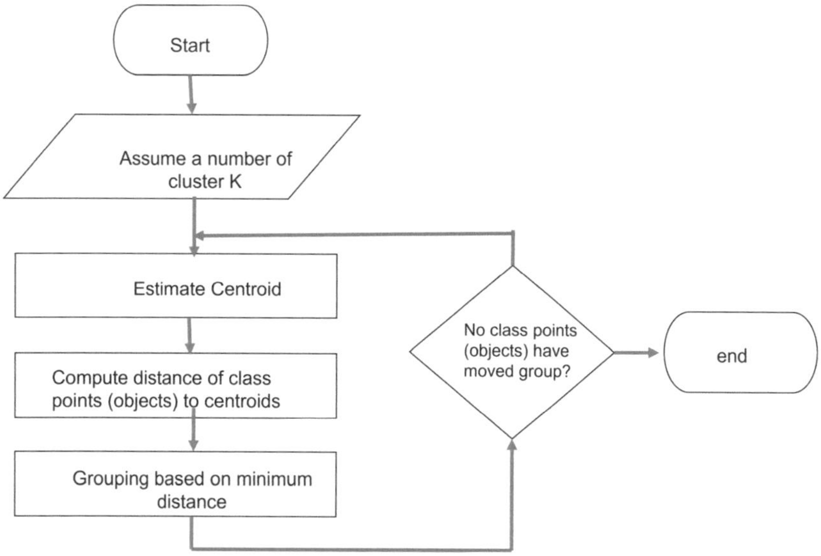


**Figure 2.**
*K-means clustering algorithm flowchart –recursive steps (adapted by [25]).*

We can re-write the above equation for points $p_n$, P is the vector of all the points where n = 1 ... M, m is the index of cluster points and M is the total number of points, and centroid point $\tau_k$ where k = 1 .... .C is the index of centroid points, C is the total number of centroid points and T is the vector which contains all the centroid points:

$$d(p_n, \tau_k) = \sqrt{\left(\tau_{x_k} - p_{x_n}\right)^2 + \left(\tau_{y_k} - p_{y_n}\right)^2} \quad (2)$$

Then each cluster point $p_n$ is assigned to a cluster based on estimating the minimum of the distance *argmindist*() to a centroid $\tau_k$ which is given by:

$$\arg\min_{\tau_k \in T} \operatorname{dist}(\tau_k, p_n)^2 \quad (3)$$

Then, we can compute the new centroid $\tau_k$ from the clustered group of points by the equation:

$$\tau_k = \frac{1}{|S_n|} \sum_{p_n \epsilon S_n} p_n \quad (4)$$

where $S_n$ is the set of all 2D data points assigned to the $k_{th}$ cluster.

Assume now we have two points in X-Y-Z three-dimensional (3D) space. Then, the Euclidean Distance between those two 3D points $p_{3D_1}(x_1, y_1, z_1)$ and $p_{3D_2}(x_2, y_3, z_2)$ is given by:

$$d\left(p_{3D_1}, p_{3D_2}\right)_{3D} = \sqrt{(x_2 - x_1)^2 + (y_2 - y_1)^2 + (z_2 - z_1)^2} \quad (5)$$

We can re-write the above Eq. (2) for a 3D centroid point $\tau_{k_{3D}}$ where k = 1 .... .C is the index of 3D centroid points, C is the total number of 3D centroid points and $T_{3D}$ is the vector which contains all the 3D centroid points:

$$d\left(p_{3D_n}, \tau_{3D_k}\right)_{3D} = \sqrt{\left(\tau_{x_k}^{3D} - p_{x_k}^{3D}\right)^2 + \left(\tau_{y_k}^{3D} - p_{y_k}^{3D}\right)^2} \quad (6)$$

where $p_{3D_n}$ are the 3D cluster points, $P_{3D}$ is the vector of all the 3D points where n = 1 ... M, n is the index of 3D cluster points and M is the total number of 3D points, and centroid point $\tau_{3D_k}$ where k = 1 .... .C is the index of 3D centroid points, C is the total number of centroid points and $T_{3D}$ is the vector which contains all the 3D centroid points.

Then, Eq. (3) can be re-written for each 3D data point and for estimating the minimum of the distance *argmin*$\text{dist}_{3D}$() to a centroid $\tau_{3D_k}$ to a centroid as follows:

$$\arg\min_{\tau_{3D_k} \in T_{3D}} \operatorname{dist}\left(\tau_{3D_k}, p_{3D_n}\right)_{3D}^2 \quad (7)$$

Then, Eq. (4) can be re-written for a 3D centroid as:

$$\tau_{3D_k} = \frac{1}{|S_{3D_n}|} \sum_{p_{3D_n} \epsilon S_{3D_n}} p_{3D_n} \quad (8)$$

where $S_{3D_n}$ is the set of all 3D data points assigned to the $k_{th}$ cluster.

## 4. Optical correlator classifiers

Since the pioneering work made by VanderLugt on spatial filtering [26–28], it became possible to construct complex matched filters. Several correlation filters for object recognition have been proposed to improve recognition capability, mostly by modifications of the amplitude or phase of the original matched filter.

### 4.1 Optical correlators categories

All the optical correlator classifiers can be further categorised based on the computational domain i.e. (a) spatial domain, (b) frequency/Fourier domain, and (c) hybrid domain by a combination of spatial and frequency domains.

For the frequency domain, correlator type of filters are commonly used. The correlator type of filters can be further classified into two main classes, the single type of filters and the cascaded type of filters. From the first class, Jamal-Aldin et al. [29–31] have presented previously their work on the non-linear difference-of-Gaussians synthetic discriminant function (NL-DOG SDF) filter. Their design of the filter was motivated by the good detectability of the modified difference of Gaussians (DOG) filter [32]. In order to improve the interclass discrimination but still keep an intraclass tolerance for a higher distortion range of the true-class object, a non-linear operation was integrated into the synthesis of the modified synthetic discriminant function filter. The DOG function approximates the second- differential operator on the image-intensity function. In practice, when convolved with the image, the DOG filter results to an edge map of the reduced-resolution image. By properly adjusting the ratio of the standard deviations of the inhibitory and excitatory Gaussians to be equal to 1.6, the DOG filter provides smoother performance for the true-class object distortions, in effect improving the intraclass properties of the filter. The NL-DOG SDF filter is based on integrating the NL-DOG operation into the synthesis of the SDF filter as well as the input test images. Mahalanobis et al. [33], were first to propose the minimum average correlation energy (MACE) filter. The MACE filter belongs to the linear combinatorial type of filters derived from the synthetic discriminant function. It is designed to maximise the training set images peak height and minimise the response of the filter to non-training set input images, with the constraint of keeping the peak-amplitude response of the filter to a fixed value for all the true-class objects included in the training set. It can be designed to give a fixed peak-amplitude response to the non-training set images, too. The solution to this resulted optimization problem is found by applying the Langrange multipliers method. The resulting MACE filter produces a sharp peak response with narrow sidelobes and with a fixed peak-height for the true-class object images included in the training set of the filter. Later, Mahalanobis et al. [34] observed that filters may perform better if hard constraints are not imposed on the correlation peaks, and suggested the use of unconstrained correlation filters. Despite the previous work in SDF synthesis that assumed the correlation values at the origin are pre-specified, there is no need for such a constraint. Thus by removing the hard constraints, we increase the number of possible solutions, thus improving the chances of finding a filter with better performance. A statistical approach is used for the design of an unconstrained filter. This method produces sharp peaks, it is computationally simpler and the proposed filters offer improved distortion tolerance. The reason lies in the fact that we do not treat training images as deterministic representations of the objects but as samples of a class whose characteristic parameters are used in encoding the filter. Three types of metrics [33, 35] are used in the design of the unconstrained filters, namely: the average correlation energy (ACE), the average similarity measure (ASM) and average correlation height (ACH). If a filter is designed to maximise the ACH criterion, it is called the maximum average correlation height (MACH) filter [34, 36]. The MACH filter maximises the relative height of the average correlation peak with respect to the expected distortions. The MACH filter yields a high correlation peak in response to the average of the training image vector. Besides optimising the ACH criterion, in practice some other performance measures, e.g. the ACE and ONV, also need to be balanced to better suit different application scenarios. Thus, based on Refregier's approach on optimal trade-off [36] filters, Mahalanobis et al. designed the optimal-tradeoff [37, 38] maximum average correlation height (OT-MACH) filter, which minimises the average correlation height criterion, holding the others constant. By adjusting the values of the three non-negative parameters of $\alpha,\beta$ *and* $\gamma$ $(0 \leq \alpha,\beta,\gamma \leq 1)$, we control the OT-MACH filter's behaviour to match different application requirements. If b = g = 0, the resulting filter behaves much like a minimum variance synthetic discriminant function (MVSDF) [39] filter with relatively good noise tolerance but broad peaks. If $\alpha = \gamma = 0$ then the filter behaves more like a MACE filter, which generally exhibits sharp peaks and good clutter suppression but is very sensitive to distortion of the target object. If $\alpha = \beta = 0$, the filter gives high tolerance for distortion but is less discriminating.

From the second class category of cascaded filters [40], Reed and Coupland [41] have studied a cascade of linear shift invariant processing modules (correlators), each augmented with a non-linear threshold as a means to increase the performance of high speed optical pattern recognition. They propose that their cascaded correlators configuration can be considered as a special case of multilayer feed-forward neural networks. They have proven that their cascaded correlator's non-linear performance can exceed the MACE filter's performance. Mahalanobis et al. [42, 43] have developed the Distance Classifier Correlation Filter (DCCF). Similarly with the work of Reed and Coupland [41], DCCF uses a cascade of shift-invariant linear filters (correlators) to compute the linear distances between the input test image and the trainset images under an optimum transformation. DCCF can be extended to support recognition of multiple object classes. Alkanhal and Kumar [44] have developed the Polynomial Distance Classifier Correlator filter (PDCCF). The underlying theory extends the original linear distance classifier correlation filter to include non-linear functions of the input pattern. PDCCF can optimise jointly all the correlators of the cascaded design, and can support multi-class object recognition.

## 4.2 Synthetic discriminant function filter

The main idea behind the Synthetic Discriminant Function (SDF) filter is to include the expected distortions in the filter design such that improved immunity to such distortions is achieved. For example, the inclusion of the out-of-class objects in the filter design achieves multi-class discrimination filter ability. In the conventional SDF filter [28] design the weighted versions of the target object are linearly superimposed, such that when the composite image is cross-correlated with any input training image, the resulting cross-correlation outputs at the origin of these cross-correlations are the same and are equal to a pre-specified constant.

The basic filter's equation constructed by the weighted combination of the training set images is:

$$h(x,y) = \sum_{i=1}^{N} a_i \cdot t_i(x,y) \tag{9}$$

where

$$a = R^{-1}c \tag{10}$$

are the weights, c is an appropriate external vector and

$$R = \iint t_i(x,y)t_j(x,y)\mathrm{dxdy} \tag{11}$$

is the correlation matrix of the training image set $t_i$.

### 4.3 Pure SDF correlator classifier for endangered species 1 and 2 speciation

**Figure 3** shows the block diagram of the Pure SDF Correlator Classifier. The train set consists of images of endangered bird species 1 and 2. Endangered bird species 1's peak value is constrained to be 0.2, and endangered bird species 2's peak value is constrained to be 1.0. By linearly superimposing the constraints weighted training set images, the composite image of the Pure SDF Classifier tool is synthesised. The test set consists of images (snags) of birds captured during an aerial survey. Each test image is then correlated with the composite image of the Pure SDF Correlator Classifier. The center peak of the output correlation plane for each input test image is then used for classifying the object snag as being either an endangered bird species 1 or an endangered bird species 2. A scatter plot of the classified endangered bird species is then drawn. For each input snag the spectral absolute peak (SAP) value Red and SAP value Blue values are used in the scatter plot.

## 5. Cognitive object recognition architecture

For the endangered species 1 and 2, we represent an object as a vector of an input image histogram's SAP for Red and Blue components. In effect, we assume that each input image of a species bird captured during an aerial survey can have a spectral signature which consists of those SAP Red and SAP Blue values. It is important to note here that K-means clustering algorithm is an unsupervised learning method. In effect, there is no a-priori information regarding the clusters size and final positions of the centroids. As we have assumed that each object is represented by two vector components, one SAP Red and one SAP Blue component, then we have a 2-Dimensional (2D) object space.

### 5.1 Biologically-inspired hybrid digital-optical system

We need to develop a new method to improve the speciation of the endangered bird species 1 and 2 for automating the image analysis from the collected datasets of the aerial surveys. There is an increased level of difficulty in correctly classifying and performing speciation for endangered bird species 1 and 2 during the winter aerial surveys due to the higher similarity of the birds' plumage between species 1 and species 2. Therefore, we propose the design and development of a novel biologically-inspired hybrid digital-optical system [14] for increasing the accuracy of the bird

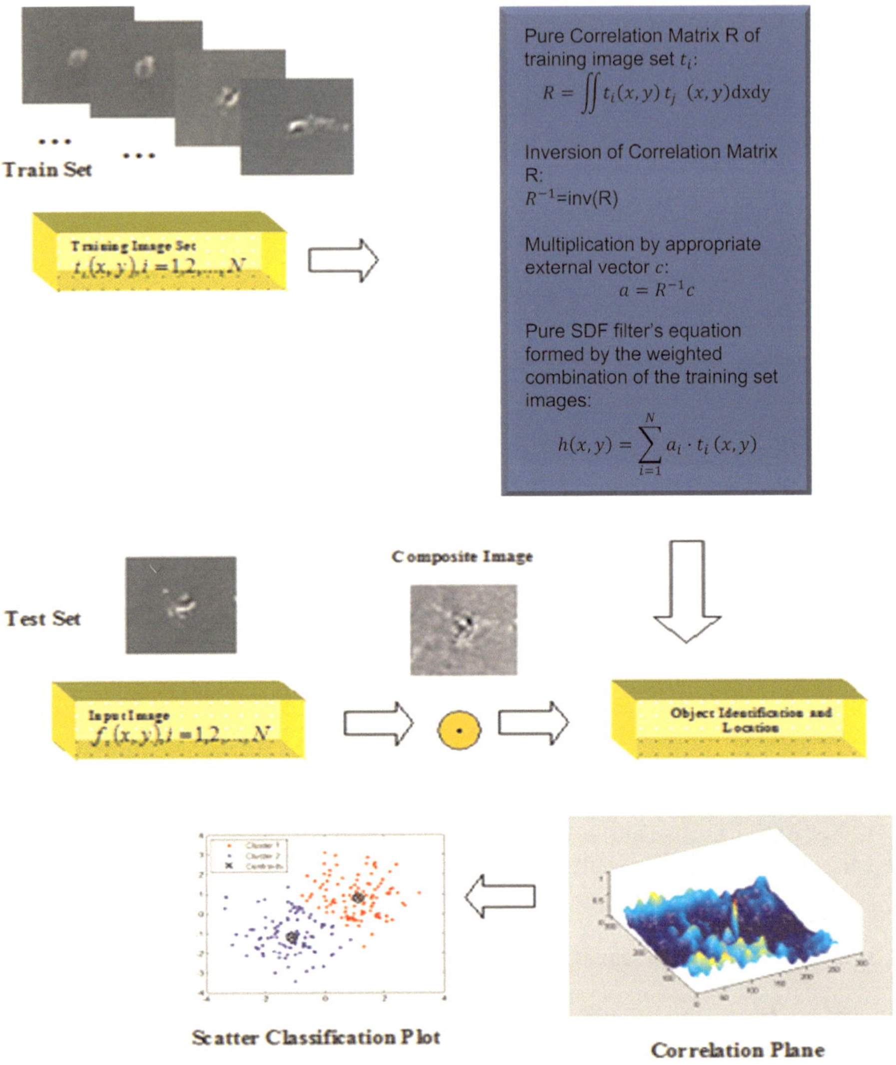


**Figure 3.**
*Pure SDF correlator classifier for endangered bird species speciation.*

species speciation and the overall time it takes to process an aerial survey. As we are going to describe in the next sections, our proposed system is capable of performing both knowledge representation and knowledge learning by incorporating in its data: (i) the shape of each endangered species 1 and 2, (ii) the size of each endangered species 1 and 2, and (iii) the colour information of each endangered species 1 and 2.

## 5.2 Knowledge representation and knowledge learning

**Figure 4** shows the block diagram of our proposed novel biologically-inspired hybrid digital-optical system called, Fast SDF K-means. It is a hybrid digital-optical design. Thus, the optical part and the K-means Clustering unit forms the digital part. The term "Fast" originates from the optical unit of the design where it consists of a correlator which can be implemented as a space domain function in a joint transform

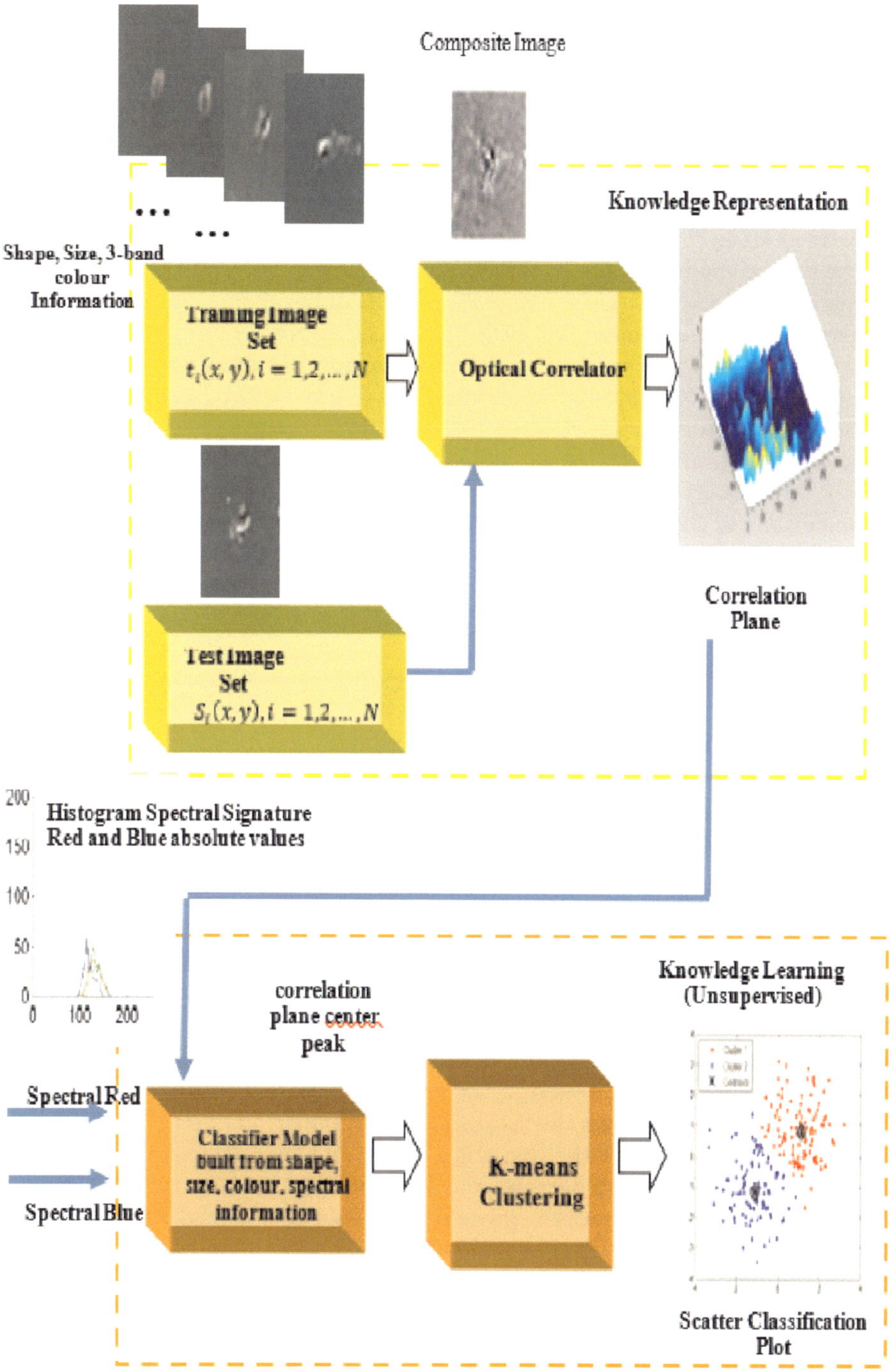


**Figure 4.**
*Fast SDF K-means classifier for endangered species speciation.*

correlator architecture or be Fourier Transformed (FT) and used as a Fourier domain filter in a 4-f Vander Lugt type optical correlator [26]. Therefore, it can operate to the speed of the light wavelength.

On **Figure 4** two different modules are shown. The first is the knowledge representation module [45, 46] which consists of the optical correlator unit, and the second is the knowledge learning module. In effect, in the first module of the knowledge representation the shape, size and 3-band colour information of each input image are synthesised into the composite image of the Fast SDF K-means Classifier. For each input object then a correlation peak value is recorded which translates this information into a numerical value. This essentially forms the knowledge representation of all the objects space in the training set.

In the second module of the knowledge learning [47], spectral histogram information together with the composite image which consists of shape, size and colour information are learned by the Fast SDF K-means Classifier. Thus, the correlation peak value together with the Red and Blue components of the spectral histogram form a 3D vector for each input object. Then, those 3D vectors which have coded the shape, size, colour and histogram information of each object are unsupervised learned by the K-means clustering unit. The output values of the Fast SDF K-means Classifier can be visualised by a scatter plot of the recognised endangered species 1 and 2 where the separate clusters of the two classes can be observed together shown with any input objects either from endangered species 1 class or from endangered species 2 class which have been misclassified.

## 6. Results

In this section, the datasets used will be described. Then, the details of the recorded results will be shown together with the Fast SDF K-means performance metrics for the different datasets.

### 6.1 Datasets

Two different datasets have been used for testing the performance of the Fast SDF K-means classifier: (a) Winter Dataset which consisted of aeriial images taken during the Winter months, and (b) Summer Dataset which consisted of aerial images taken during the Summer months. Winter Dataset consisted of 221 three-band JPEG formatted aerial image shots also known as snags. Each snag had the size of [320240] pixels. Summer Dataset consisted of 270 three-band JPEG formatted snags. Each snag had, as for the Winter Dataset, the size of [320240] pixels. For aerial survey logging and identification reasons, all the shapefiles ".jgw" and ".mat" tagged information files have been saved for both datasets, too. To match the aerial survey automated image analysis of the snags with the ground truth data an object identification (ObjectID) information had been provided with each snag. Ground truth data had been collected from sea surveys or on-shore remote view including the total number of endangered species 1 and the total number of endangered species 2.

#### *6.1.1 Winter dataset results*

**Figure 5** shows the K-means clustering algorithm classification scatter plot. The classified endangered species 1 and species 2 are drawn against their SAP Red (x-axis) and SAP Blue (y-axis) values. All the objects snags were classified using their histogram spectral values of SAP Red and SAP Blue. There is a high deviation and population ratio reverse between the circa ratio of species 1 and species 2 given by the boat

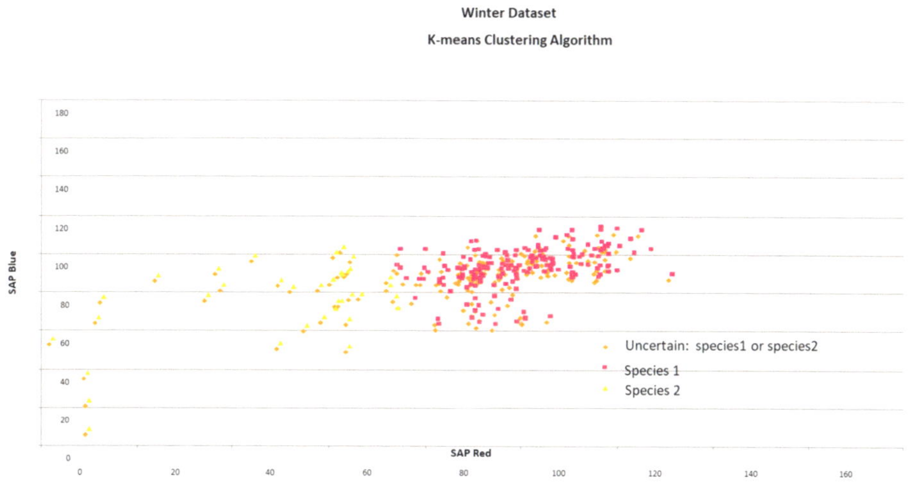


**Figure 5.**
*Winter dataset speciation: K-means clustering algorithm. It shows the scatter plot of the classified endangered species 1 and species 2. The classified endangered species are drawn against their SAP red (x-axis) and SAP blue (y-axis) values. There is a high number of endangered species which their ID tagged by the GT scientists was uncertain i.e. tagged as* "species 1 or species 2".

survey i.e. *(species1/species2)* = *13:1* and the classified by the algorithm endangered species i.e. classified *species2* = *185* and classified *species1* = *36*.

**Figure 6** shows the Pure SDF Correlator classification scatter plot. The classified endangered species 1 and species 2 are drawn against their SAP Red (x-axis) and SAP Blue (y-axis) values. However, now all the objects snags were classified using their shape, size and 3-band colour information. This time the classified by the Pure SDF Correlator endangered species produced a ratio of *(species1/species2)* = *24:2*.

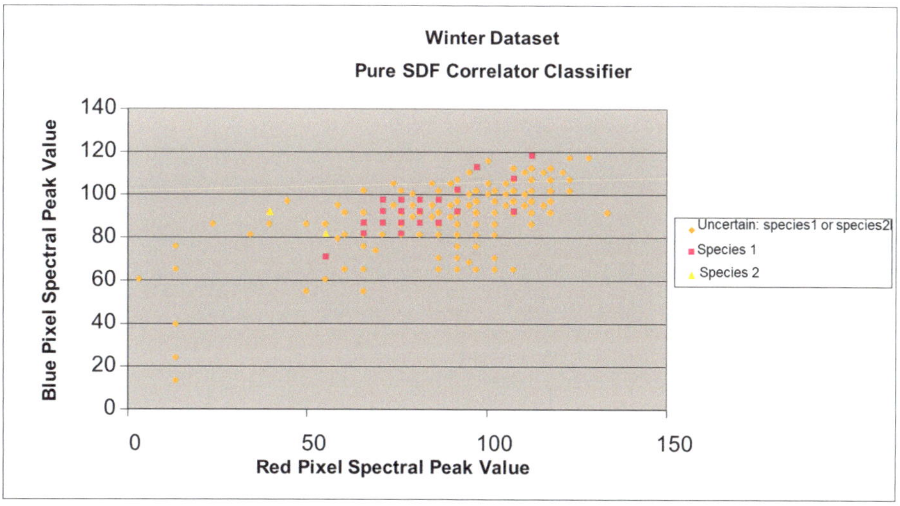


**Figure 6.**
*Winter dataset speciation: Pure SDF correlator classifier. It shows the scatter plot of the classified endangered species 1 and species 2. All the objects snags were classified using the correlation peak value of each input image which encoded their shape, size and colour information. The classified endangered species are drawn against their SAP red (x-axis) and SAP blue (y-axis) values. There is a high number of endangered species which their ID tagged by the GT scientists was uncertain i.e. shown on the plot* as "species 1 or species 2".

**Figure 7** shows the Fast SDF K-means Classifier scatter plot. The classified endangered species 1 and species 2 are drawn against their SAP Red (x-axis) and SAP Blue (y-axis) values. Now all the objects snags were classified using their 3D vectors which encode shape, size, 3-band colour, and spectral histogram information. This time the classified by the Fast SDF K-means Classifier produced a ratio of *(species1/species2) = 20:4*.

### 6.1.2 Summer dataset results

**Figure 8** shows the K-means Clustering algorithm classification scatter plot. The classified endangered species 1 and species 2 are drawn against their SAP Red (x-axis) and SAP Blue (y-axis) values. All the objects snags were classified using their histogram spectral values of SAP Red and SAP Blue. There is a high deviation and population ratio reverse between the circa ratio of species 1 and species 2 given by the boat survey i.e. *(species1/species2) = 15:2* and the classified by the algorithm endangered species i.e. classified *species2 = 151* and classified *species1 = 96*.

**Figure 9** shows the Pure SDF Correlator classification scatter plot. The classified endangered species 1 and species 2 are drawn against their SAP Red (x-axis) and SAP Blue (y-axis) values. However, now all the objects snags were classified using their shape, size and 3-band colour information. This time the classified by the Pure SDF Correlator endangered species produced a ratio of *(species1/species2) = 11:1*.

**Figure 10** shows the Fast SDF K-means Classifier scatter plot. The classified endangered species 1 and species 2 are drawn against their SAP Red (x-axis) and SAP

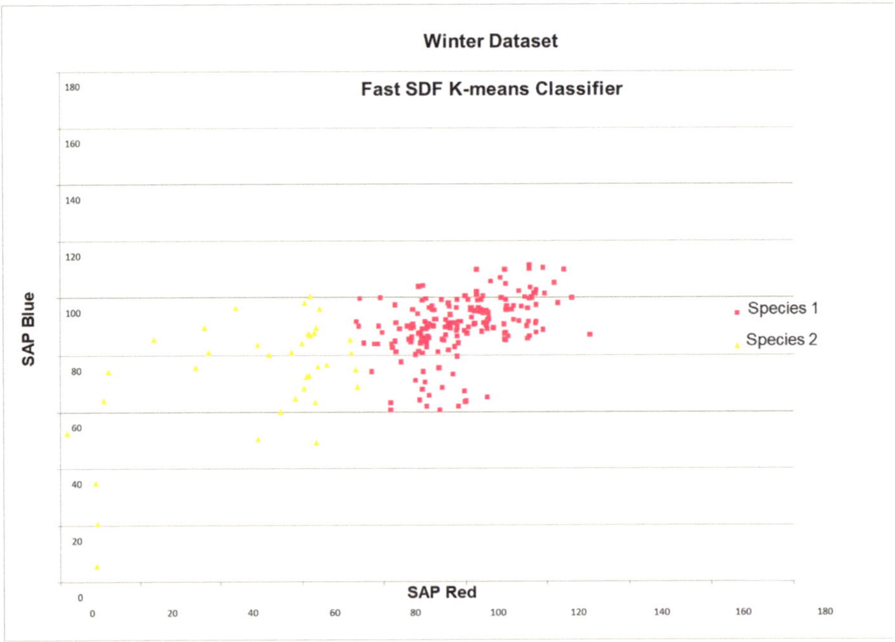


**Figure 7.**
*Winter dataset speciation: Fast SDF K-means classifier. It shows the scatter plot of the classified endangered species 1 and species 2. All the objects snags were classified using their 3D vectors which have encoded their shape, size, 3-band colour and histogram spectral information. The classified endangered species are drawn against their SAP red (x-axis) and SAP blue (y-axis) values.*

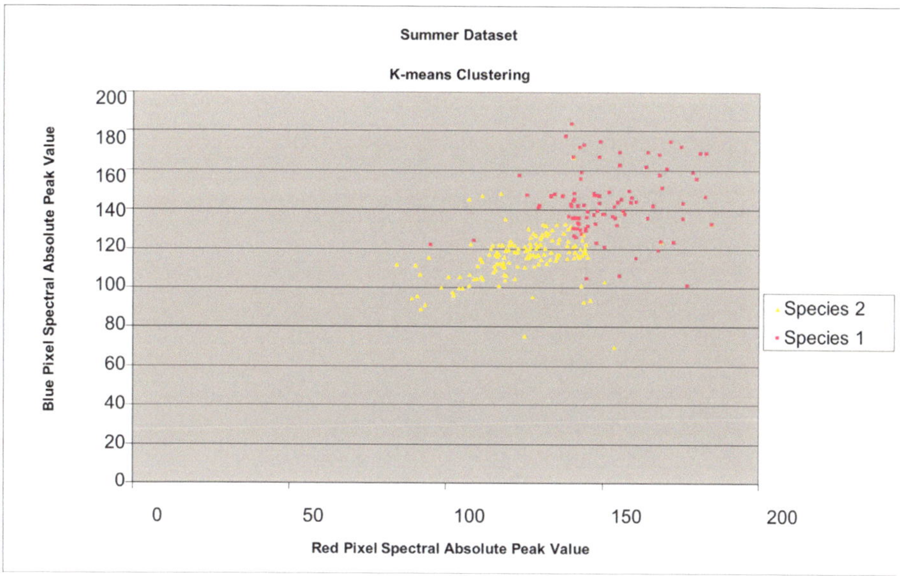


**Figure 8.**
*Summer dataset speciation: K-means clustering algorithm. It shows the scatter plot of the classified endangered species 1 and species 2. The classified endangered species are drawn against their SAP red (x-axis) and SAP blue (y-axis) values.*

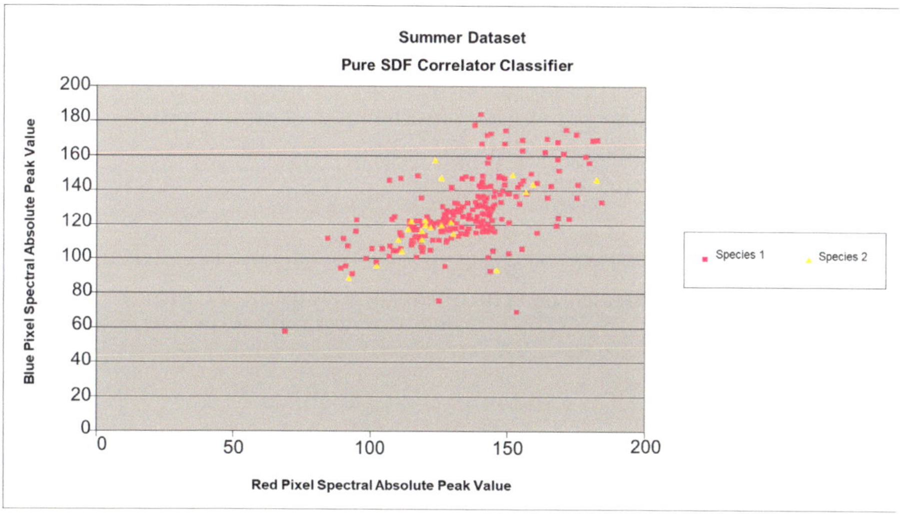


**Figure 9.**
*Summer dataset speciation: Pure SDF correlator classifier. It shows the scatter plot of the classified endangered species 1 and species 2. All the objects snags were classified using the correlation peak value of each input image which encoded their shape, size and colour information. The classified endangered species are drawn against their SAP red (x-axis) and SAP blue (y-axis) values.*

Blue (y-axis) values. Now all the objects snags were classified using their 3D vectors which encode shape, size, 3-band colour, and spectral histogram information. This time the classified by the Fast SDF K-means Classifier produced a ratio of *(species1/species2) = 15:1.*

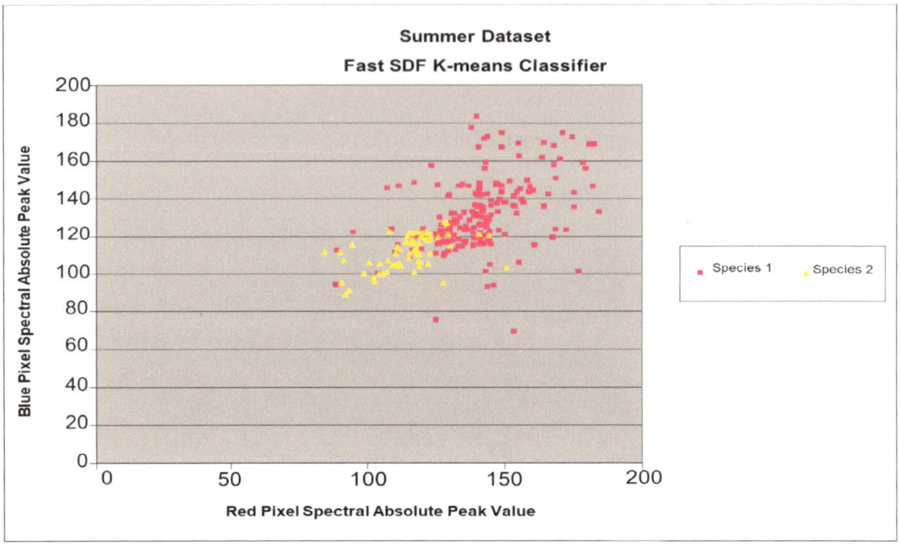


**Figure 10.**
*Summer dataset speciation: Fast SDF K-means classifier. It shows the scatter plot of the classified endangered species 1 and species 2. All the objects snags were classified using their 3D vectors which have encoded their shape, size, 3-band colour and histogram spectral information. The classified endangered species are drawn against their SAP red (x-axis) and SAP blue (y-axis) values.*

### *6.1.3 Performance metrics*

Performance metrics have been used to assess the k-means classification algorithm, the Pure SDF correlator classifier, and the novel Fast SDF k-means classifier. Thus, precision metric is given by the ratio of true positives (TP) over the total number of false positives (FP) plus true positives:

$$\text{Precision} = \frac{\text{TP}}{\text{TP} + \text{FP}} \tag{12}$$

Recall metric is computed as the ratio of the TP versus the TP plus the false negatives (FN):

$$\text{Recall} = \frac{\text{TP}}{\text{TP} + \text{FN}} \tag{13}$$

True negative rate (TNR) is computed as the ratio of the true negatives (TN) versus the TN plus the FP:

$$\text{TNR} = \frac{\text{TN}}{\text{TN} + \text{FP}} \tag{14}$$

Accuracy is given by the ratio of TP plus TN over the sum of TP, TN, FP and FN:

$$\text{Accuracy} = \frac{\text{TP} + \text{TN}}{\text{TP} + \text{TN} + \text{FP} + \text{FN}} \tag{15}$$

**Table 1** shows the performance metric values for the winter dataset of all the three tested classifiers. In effect, the second column shows the performance metric values

| | Winter Speciation | | |
|---|---|---|---|
| | **k-means clustering** | **Pure SDF correlator classifier** | **Fast SDF k-means classifier** |
| Precision % | 16 | 86.33 | 96.11 |
| Accuracy % | 4 | 84.33 | 89.6 |
| TNR % | 0 | 4.3 | 0 |
| Recall % | 51.7 | 97.2 | 92.96 |

**Table 1.**
*Winter dataset speciation performance metric values for the k-means clustering algorithm, the pure SDF correlator and the fast SDF k-means classifier.*

| | Summer Speciation | | |
|---|---|---|---|
| | **k-means clustering** | **Pure SDF correlator classifier** | **Fast SDF k-means classifier** |
| Precision % | 42.5 | 83.94 | 89.93 |
| Accuracy % | 31.87 | 82.0 | 83.23 |
| TNR % | 0 | 15.0 | 0 |
| Recall % | 56.04 | 97.2 | 91.80 |

**Table 2.**
*Summer dataset speciation performance metric values for the k-means lustering algorithm, the pure SDF correlator and the fast SDF k-means classifier.*

of the k-means classification algorithm, the third column shows the performance metric values of the Pure SDF correlator classifier, and the fourth column shows the performance metric values of the novel Fast SDF k-means classifier. Similarly, **Table 2** shows the performance metric values for the summer dataset of all the three evaluated classifiers. The results are shown on the corresponding columns of **Table 2** as for **Table 1**.

## 7. Discussion and conclusions

In this section, the results of our evaluated classifiers are analysed and compared. In particular, the discussion is focused on the performance of the three classifiers during the winter dataset tests since the endangered *species 1* and *species 2* offer a greater classification challenge during the winter months than the summer ones. A separate section is included with the main conclusions of this work together with some future research suggestions.

### 7.1 Discussion

From **Figures 5–7** it can be found that, overall, the Fast SDF k-means has performed better than the k-mean classification algorithm and the Pure SDF correlator classifier, too, for the winter dataset. Though the Pure SDF correlator classifier gave a classification rate of *(species1/species2)* closer to the ground-truth (GT) scientists ratio than the Fast SDF k-means, it still had a much higher, almost the double of the total number of classified endangered species i.e. sum of species 1 and species 2, of uncertain classifications which cannot be matched with either species1 or species2. Similarly, from **Figures 8–10** it can be found that, overall, the Fast SDF k-means has performed better than the k-mean classification algorithm and the Pure SDF correlator classifier, too, for the summer dataset. Thus, the classification rate of *(species1/species2)* for the Fast SDF k-means was almost identical to the classification rate of the GT scientists. By incorporating the shape, size, 3-band colour and histogram spectral information into the Fast SDF k-means classifier it has improved the classification performance for both summer and winter datasets in comparison to the other two classifiers.

From **Tables 1** and **2** the performance of all the classifiers can be assessed which they were used for the winter endangered bird species plumage and summer endangered bird species plumage speciation. It is can be clearly shown that k-means classification algorithm performed worse than the Pure SDF Correlator and Fast SDF K-means classifiers for both summer and winter endangered bird species plumage. In effect, the precision value was 96.11% and the accuracy value was 89.6% for the Fast SDF k-means classifier when tested with the winter dataset. The precision value became 89.93% and the accuracy value reached 83.23% for the Fast SDF k-means classifier when tested with the summer dataset. It worth of mentioning that the precision values of the Fast SDF k-means classifier for both datasets were higher than the human image analysts values which was not more than 88% for the winter plumage and not more than 89% for the summer plumage.

It should be noted that during the summer and winter surveys the weather conditions were significantly different. In effect, during the winter aerial survey the weather conditions were poor but during the summer boat survey the conditions were significantly improved. Thus, by observing the performance metric values of **Tables 1** and **2**, it can be concluded that the Pure SDF Correlator Classifier's performance has not been significantly affected due to the different weather conditions when the surveys were conducted. Also, though the performance of Fast SDF k-means classifier seems to deviate on summer survey from the winter survey this was found to be due to glint effects in the capture image data. After examining the summer datasets, it was identified approximately 25% of the total number of snags to have significant glint effects in them. Nevertheless, the overall performance of the Pure SDF correlator classifier and the Fast SDF k-means classifier closely matched the boat survey during both winter and summer weather conditions.

Further, the novel Fast SDF k-means classifier has minimised the amount of total data needed to be ground-truthed by the GT scientists i.e. it can lead to an increased automation of the speciation process. We assessed the Fast SDF k-means classifier precision and accuracy values to be greater than 85% during the winter surveys i.e. approximately only 20% or less of the total amount of survey data would need to be ground-truthed. Hence, that would make more cost-effective the processing of the datasets i.e. more surveys per day would become possible to be processed by the GT scientists.

## 7.2 Conclusion

We have shown how our novel cognitive architecture of the Fast SDF k-means classifier is conceptually connected with the image analysis processing cycle. It combines a hybrid digital-optical design where the k-means unsupervised learning algorithm is integrated with a correlator. Thus, Fast SDF k-means classifier consists of a knowledge representation module formed by the SDF correlator and a knowledge learning module formed by the k-means classifier. The shape, size and 3-band colour information of each input image is synthesised into the composite image of the Fast SDF k-means classifier and, then, a corresponding correlation value is recorded which translates this information into a numerical value. Then, the knowledge learning module formed by the k-means classifier will learn the coded 3D vector of the composite image together with the Red and Blue components of the spectral histogram for each input object.

We have assessed the novel Fast SDF k-means classifier using performance metrics, and, then, compared it with the k-means classifier and the Pure SDF correlator classifier, too. The k-means classifier learns the vectors of the SAP Red (x-axis) and SAP Blue (y-axis) values of the input dataset. The Pure SDF correlator classifier uses the correlation peak value of each input image which encoded their shape, size and colour information. The precision values of the Fast SDF k-means classifier for both datasets were found to be higher than the human image analysts values which was not more than 88% for the winter plumage and not more than 89% for the summer plumage. Both the Pure SDF correlator classifier and the Fast SDF k-means classifier consistently performed under the different summer and winter conditions. Though the other types of surveys, e.g. boat survey can be prone to human error, applying those performance metrics with our developed Fast SDF k-means correlator classifier could be used for quality control (QC) and quality assessment (QA) of the classifier's results over the aerial survey data.

In Section 4.1, NL-DoG SDF correlator classifier was described. NL-DoG SDF correlator in comparison with the Pure SDF correlator offers improved detectability and interclass discrimination but still keep an intraclass tolerance for a higher distortion range of the true-class object. Thus, we propose in future work to integrate a NL-DoG in our Fast SDF k-means classifier design which is expected to enhance its performance.

## Author details

Ioannis Kypraios
School of Computer Science and Informatics, Institute of Artificial Intelligence, Faculty of Computing, Engineering and Media, De Montfort University, Leicester, UK

*Address all correspondence to: kypraios.ioannis@gmail.com

## References


[1] Historic England. Heritage and the Economy [Internet]. 2020. Available from: https://historicengland.org.uk/content/heritage-counts/pub/2020/heritage-and-the-economy-2020/ [Accessed: 26 October 2022]

[2] Kordej-De VŽ, Cultural ŠI. Heritage, tourism and the UN sustainable development goals: The case of Croatia. In: Andreucci MB, Marvuglia A, Baltov M, Hansen P, editors. Rethinking Sustainability Towards a Regenerative. Vol. 15. Economy: Future City: Springer; 2021

[3] Heritage Lottery Fund (HLF). Investing in Success: Heritage and the UK Tourism Economy [Internet]. 2010. Available from: https://www.heritagefund.org.uk/sites/default/files/media/about_us/hlf_tourism_impact_single.pdf [Accessed: 26 October 2022]

[4] Barbosa AEA, Tella JL. How much does it cost to save a species from extinction? Costs and rewards of conserving the Lear's macaw. Royal Society open science. 2019;**6**:190190

[5] Chen F, Guo H, Tapete D, Masini N, Cigna F, Lasaponara R, et al. Interdisciplinary approaches based on imaging radar enable cutting-edge cultural heritage applications. National Science Review. 2021;**8**(9)

[6] Kaur J, Singh W. Tools, techniques, datasets and application areas for object detection in an image: A review. Multimedia Tools Applications. 2022;**81**: 38297-38351

[7] Lo CP. Modern use of aerial photographs in geographical research. JSTOR Area. 1971;**3**(3):164-169

[8] Haykin S. Neural Networks and Learning Machines. 3rd ed. New York: Pearson; 2009

[9] Serre T, Oliva A, Poggio T. A feedforward architecture accounts for rapid categorization. PNAS Biological Sciences. 2007;**104**(15): 6424-6429

[10] DiCarlo JJ, Zoccolan D, Rust NC. How does the brain solve visual object recognition? Neuron. 2012;**73**(3): 415-434

[11] Kriegeskorte N, Douglas PK. Cognitive computational neuroscience. Nature Neuroscience. 2018;**21**(9): 1148-1160

[12] Lee I, Portier B. An empirical study of knowledge representation and learning within conceptual spaces for intelligent agents. In: Proceedings of the IEEE/ACIS International Conference Computer and Information Science; 11–13 July 2007; Melbourne. Australia: IEEE; 2007. pp. 463-468

[13] Aler R, Borrajo D, Isasi P. Knowledge representation issues in control knowledge learning. In: Proceedings of the 17th International Conference on Machine Learning (ICML'00). San Francisco, CA. USA; 2000. p. 1-8. ISBN 1558607072

[14] Kypraios I. Performance analysis of the modified-hybrid optical neural network object recognition system within cluttered scenes. In: Kypraios I, editor. Advances in Object Recognition Systems. London, UK, London, UK: InTech; 2012 ISBN: 978-953-51-0598-5

[15] Bisang W. Knowledge representation and cognitive skills in problem solving. In: Zlatkin-Troitschanskaia O, Wittum G, Dengel A, editors. Positive Learning in the Age of Information. Wiesnaden: Springer V. S; 2018

[16] A Tutorial on Clustering Algorithms [Internet]. 2012. Available from: http:// home.dei.polimi.it/matteucc/Clustering/ tutorial_html/kmeans.html [Accessed: 26 October 2022]

[17] Russell S, Norvig P. Artificial Intelligence: A Modern Approach. 3rd ed. New York: Prentice Hall; 2009. p. 816-820. ch20

[18] MacQueen JB. Some methods for classification and analysis of multivariate observations. In: Proceedings of the 5th Berkeley Symposium on Mathematical Statistics and Probability. Vol. 1. Berkeley: University of California; 1967. pp. 281-297

[19] Bennet J, Ganaprakasam CA, Arputharaj K. A discrete wavelet based feature extraction and hybrid classification technique for microarray data analysis. The Scientific World Journal. 2014;**2014**:1-9. Article ID 195470. DOI: 10.1155/2014/195470

[20] Wu S, Chen H, Zhao Z, Long H, Song C. An improved remote sensing image classification based on k-means using HSV colour feature. In: Proceedings of the 10th International Conference on Computational Intelligence and Security (CIS-2014). Kumming, Yunnan. China; 15-16 November 2014. p. 201-204

[21] Abbas A, Minallh N, Ahmad N, Abid SA, R, Khan M. A. A. K-means and ISODATA clustering algorithms for landcover classification using remote sensing. Sindh University Research Journal. 2016;**48**

[22] Vishwanath N, Ramesh B, Rao SP. Unsupervised classification of remote sensing images using k-means algorithm. International Journal of Latest Trends in Engineering and technology (IJLTET). 2016;7(2)

[23] Yin X, Sasaki Y, Wang W, Shimizu K. YOLO and k-Means Based 3D Object Detection Method on Image and Point Cloud [Internet]. 2020. Available from: https://arxiv.org/abs/2004.11465

[24] Nguyen H, Lee E-H, Bae C, Lee S. Multiple object detection based on clustering and deep learning methods. Sensors. 2020;**16**(4424). DOI: 10.3390/ s20164424

[25] Teknomo K. K-Means Clustering Tutorials [Internet]. Available from: http://people.revoledu.com/kardi/tutoria l/kMean/ [Accessed: 26 October 2022]

[26] Vander LA. Signal detection by complex spatial filtering. IEEE Transactions on Information Theory (IT-10). 1964;**10**(2):139-145

[27] Bahri Z, Kumar BVK. Generalized synthetic discriminant functions. Journal of Optical Society of America. 1988;**5**(4): 562-571

[28] Kumar BVK. Tutorial survey of composite filter designs for optical correlators. Applied Optics. 1992;**31**(23): 4773-4801

[29] Jamal-Aldin LS, Young ECD, Chatwin CR. Application of non-linearity to wavelet-transformed images to improve correlation filter performance. Applied Optics. 1997; **36**(35):9212-9224

[30] Jamal-Aldin LS, Young ECD, Chatwin CR. Nonlinear preprocessing operation for enhancing correlator fitler performance in clutter. In: Proceedings of the European Optical Society OC' 98 Conference on Optics in Computing. Vol. 3490. Belgium: SPIE; 1998. pp. 182-186

[31] Jamal-Aldin LS, Young ECD, Chatwin CR. In-class distortion

tolerance, out-of-class discrimination and clutter resistance of correlation filters that employ a space domain non-linearity applied to wavelet filtered input images. SPIE. 1998;**3386**:111-122

[32] Shang L, Wang RK, Chatwin CR. Frequency multiplexed DOG filter. Optics and lasers in engineering. Elsevier Applied Science. 1997;**27**(2):161-177

[33] Mahalanobis A, Vijaya Kumar BVK, Casasent D. Minimum average correlation energy filters. Applied Optics. 1987;**26**(17):3633-3640

[34] Vijaya KB, V. K, Hassebrook L. Performance measures for correlation filters. Applied Optics. 1990;**29**(20): 2997-3006

[35] Mahalanobis A, Vijaya Kumar BVK, Song S, Sims SRF, Epperson JF. Unconstrained correlation filters. Applied Optics. 1994;**33**(17):3751-3759

[36] Refregier P. Filter design for optical pattern recognition: Multicriteria optimisation approach. Optics Letters. 1990;**15**(15):854-856

[37] Zhou H, Chao TH. MACH filter synthesising for detecting targets in cluttered environment for gray-scale optical correlator. SPIE. 1999;**715**:394-398

[38] Mahalanobis A, Vijaya KB, V. K. Optimality of the maximum average correlation height filter for detection of targets in noise. Optical Engineering. 1997;**36**(10):2642-2648

[39] Vijaya KB, V. K. Minimum variance synthetic discriminant functions. Journal of Optical Society of America A. 1986;**3**: 1579-1584

[40] Dubois F. Non-linear cascaded correlation processes to improve the performances of automatic spatial-frequency-selective filters in pattern recognition. Applied Optics. 1996; **35**(23):4589-4597

[41] Reed S, Coupland J. Cascaded linear shift-invariant processors in optical pattern recognition. Applied Optics. 2001;**40**(23):3843-3849

[42] Mahalanobis A, Vijaya Kumar BVK, Sims SRF. Distance classifier correlation filters for multiclass target recognition. Applied Optics. 1996;**35**(17):3127-3133. DOI: 10.1364/AO.35.003127. PMID: 21102690.

[43] Mahalanobis A, Vijaya KB, V. K., Sim S. R. F. Distance-classifier correlation filters for multiclass target recognition. Applied Optics. 1996;**35**(17): 3127-3133

[44] Alkanhal M, Vijaya KB, V. K. Polynomial distance classifier correlation filter for pattern recognition. Applied Optics. 2003;**42**(23):4688-4708

[45] Khalifa M, Ning Shen K. Effects of knowledge representation on knowledge acquisition and problem solving. The Electronic Journal of Knowledge Management. 2006;**4**(2):153-158

[46] Policastro CA, Zuliani G, da Silva RR, Romero RAF. Hybrid knowledge representation applied to the learning of the shared attention. In: Proceedings of the International Joint Conference on Neural Networks (IJCNN-2008); IEEE World Congress on Computational Intelligence (WCCI-2008); 1–6 June 2008. Hong Kong, China: IEEE; 2008. pp. 866-870

[47] Ho SB, Liausvia F. Knowledge representation, learning, and problem solving for general intelligence. In: Proceedings of the 6th International Conference on Artificial General Intelligence. 2013. DOI: 10.1007/978-3-642-39521-5-7